The Interpretation of
QUANTUM MECHANICS

The Interpretation of
QUANTUM
MECHANICS

Michael Audi

The University of Chicago Press

Chicago and London

The University of Chicago Press, Chicago 60637
The University of Chicago Press, Ltd., London
© 1973 by The University of Chicago
All rights reserved. Published 1973
Printed in the United States of America
International Standard Book Number: 0–226–03177–2
Library of Congress Catalog Card Number: 73–78663

MICHAEL AUDI is assistant professor of philosophy
at Southern Illinois University. This is his first book.
[1973]

To Barbara and Leslie,
for devotion far beyond duty

Contents

Preface

The central thesis of this book is twofold: first, quantum mechanics is an essentially indeterministic theory and, by extension, those physical processes described by the theory are indeterministic; second, if this indeterminism is genuinely accepted, all the philosophical problems of interpreting quantum theory become tractable. Such problems arise because the theories of classical physics are not merely false in detail, when applied to phenomena requiring quantum mechanical analysis, but are based on a conceptual structure—one using wave and particle models, and presupposing determinism—which seems completely inapplicable to such phenomena. Accordingly, one is strongly tempted to infer that quantum and classical theory are conceptually incompatible in all important respects.

This conceptual incompatibility would immediately lead to nasty, if not unanswerable, questions. For instance, if the meaning of key terms common to classical and quantum theory cannot be explicated by means of classical models, what then do the terms mean in quantum theory, and how could we ever come to understand these meanings? Further, what is the relation between the two theories: how can they be compared, if they have no common terms with theory-invariant meaning? Finally, is our concept of determinism, and the related notion of causality, such that an indeterministic theory could rationally be accepted as correct?

Fortunately, the appearance of complete conceptual incompatibility is just that—an appearance, not a reality. In this study, I demonstrate that the key common terms are meaning-invariant, and establish a unitary particle interpretation of material object phenomena, and a unitary wave-field interpretation of electro-magnetic radiation. This showing can be reconstructed as a three-stage process.

First, I show that the central idea of classical mechanics is separable into two distinct, independent ideas—the claim that physical systems are particles is separable from the further classical claim that some set of laws determines the individual paths of indi-vidual particles. Further, I demonstrate that neither the Heisenberg uncertainty relations, nor diffraction and "interference" effects, nor any other quantum mechanical phenomena preclude particle analyses. Second, I show that the actual usage of the terms in question by physicists reflects meaning-invariance. Third, the pre-supposition of the first stage—that quantum theory, and, by exten-sion, the physical processes described thereby, are indeterministic—is justified. This justification depends on, among other things, a crucial distinction between causality and determinism, and an analysis of criteria for confirmation of competing theories or hypoth-eses. The executions of these stages overlap considerably. Chap-ters 2, 5, and 6 are specifically devoted to the first task, chapter 2 to the second task, and chapters 3, 4, and 5 to the third task, but all seven chapters touch on all three aspects, in some way or other.

I claim many ancillary benefits for my analysis, the first being a careful explication of the much maligned, and often misunderstood, Copenhagen interpretation. I attack this interpretation, due to Bohr and Heisenberg, in the strongest of terms—showing it to be either inconsistent or uninformative—but I attack it on its central theses, rather than by refuting disreputable philosophy—"positivism," "subjectivism," "instrumentalism," and so on—often vaguely asso-ciated with, but inessential to, the interpretation. My presentation of the Copenhagen interpretation, in chapters 1 and 2, is based on virtual paraphrase of the writings of Bohr and Heisenberg in the period (1927–29) when the interpretation was first formulated, and from which its influence dates.

Lest there be any misunderstanding, some disclaimers are in order here. I make no claims about what Bohr and Heisenberg believed prior to, say, 1925, or after, say, 1935. Contrary to the working assumptions of various learned scholars, such beliefs are irrelevant—those before 1925 because the problematic theoretical results the interpretation was primarily designed to resolve had not yet come into existence (Schrödinger's wave equation and Heisenberg's uncertainty relations); those after 1935 because the interpretation going by the name 'Copenhagen' had already begun to exert considerable influence in the early and middle thirties, and hence its essential elements should, with benefit of hindsight, be discernible by 1935, if not considerably earlier. The year 1935 is also the date of the famous Einstein, Podolsky, and Rosen challenge to the Copenhagen interpretation and of Bohr's instant reply. Careful examination of that reply reveals it to be based on exactly those tenets I ascribe to Bohr and Heisenberg in my first two chapters, tenets they held consistently from 1927 to, at least, 1935.

I do not deny that Bohr and Heisenberg were, perhaps even in the period 1927 to 1935, positivists, instrumentalists, or subjectivists. My only point is that they neither used, nor needed, nor committed themselves to, any positivist, subjectivist, or instrumentalist arguments in those texts many historians agree are the crucial primary sources. What they said in their memoirs, or the functional equivalent thereof, is interesting only to those gullible enough to take a man's word, or, worse, his hint, on how he used to argue, rather than examine actual specimens of earlier argumentation.

Another ancillary benefit is a clarification of the notion of determinism. More precisely, I isolate three separately necessary and jointly sufficient conditions for correctly labeling a theory 'deterministic,' and show that quantum mechanics does not satisfy two of these conditions. Following this, I indicate why deterministic hidden variable theories—which are distinct theoretical competitors of "standard" quantum mechanics, and not reinterpretations of it—cannot be regarded as satisfactory, on present evidence. I hope to have properly reflected the tentative character of philosophic conclusions grounded on developments in physical theory. In particular, once we realize that determinism is in the first instance a feature of theories, and only by extension a feature of those physical processes described thereby, then we perceive that, if our opinions about the deterministic or indeterministic character of the world are to be rational, we must be prepared to revise our opinions as new and better theories arise. I hope, even more strongly, that no one will suppose, as some hidden variable theorists seem to think, that present judgments should be based on something other than present evidence.

Yet other ancillary benefits include revelation of the desperate character of various alternative interpretations: dualism, revised logics, and others. Finally, in discussing relativistic and field theory effects, I enumerate the many crucial differences between photons and material particles, and expose as completely unfounded the fashionable claim of equivalence between particle and wave-field descriptions within quantum field theory.

A series of concluding remarks on attitude and approach is now in order. The reader will find in this book no discussion of the "problem of measurement" in quantum theory. This is as it should be. Only the briefest reflection is required to realize that this problem is created by adoption of an unsatisfactory interpretation of quantum mechanics. Crudely, the problem is this: if one applies a quantum mechanical description to a total system consisting of, say, a microscopic particle plus a macroscopic instrument, what seems to be predicted is that the (total) system won't be in any particular state at all, after interaction of particle and instrument—the "pointer" on a meter won't point anywhere in particular. This result should be viewed as a reductio ad absurdum of one of the two controversial premises from which it is derived: that individual processes are indeterministic; and that the quantum mechanical description—typically, solutions of Schrödinger's wave equation—applies to individual processes, rather than to statistical ensembles thereof. One of the principal morals of my story is that the second premise is indeed absurd, while the first is a sound insight of Bohr's.

In chapter 4, following my discussion of Bohm's 1952 hidden variable theories, I have attempted to show that my results are not only consonant with, but illustrative of, the upshot of recent analyses of hidden variable theories in general, in relation to von Neumann's theorem. It is essential that the reader remember the redundancy of the phrase 'deterministic hidden variable theory.' Results displaying the absurdity and/or empirical falsity of hidden variable theories are devastating blows to determinists, but irrelevant to those of us who agree with hidden variablists that physical systems have exact intrinsic properties, while disagreeing with their claim that the postinteraction values of these properties are uniquely determined by the preinteraction values. This is not a very subtle distinction, and it is probably only adepts in the field who will prove incapable of grasping it.

Even if I succeed in convincing the reader that determinism and causality are quite different concepts, he may find my characterization of the latter trivial and relatively uninformative. To say, for example, that a process, or a theoretical description thereof, is causal if and only if it satisfies, or incorporates, some versions of the laws of conservation of energy and momentum, may not seem to add much to the conversation. In reply, I can only offer the following perspective. There are two distinct theses, often conflated, which sometimes go by the name 'principle of causality': first, that all effects have causes, and, second, that identical causes produce identical effects. The second thesis—determinism—is hardly trivial. If the analyses of this book are correct, it is empirically false. The first thesis seems indubitable. It is very likely an a priori and hence, presumably, an analytic truth—true in virtue of the meanings of the terms expressing it. Like many, though by no means all, analytic truths, it may seem trivial. I am untroubled by this. It is a sound philosophic rule of thumb, a well-supported statistical generalization, that almost all true philosophical theses seem trivial and that almost all interesting ones turn out to be false.

One final methodological point deserves mention. In the various conceptual analyses of this essay, I regard my conclusions as descriptive, not prescriptive. When I claim that 'particle' means the same in both classical and quantum mechanics, I am not recommending a new usage for physicists. I am claiming that their past and present usage, interpreted in the only way which seems to make any sense, already reflects this meaning-invariance. Of course, if one asks physicists explicitly for an opinion about the alleged meaning-invariance, they may and do say anything at all. This only proves they are no more infallible than any other group, when speaking as amateur philosophers rather than as professional physicists. Only Socratic polls, not Gallup polls, are appropriate for philosophic research.

Philosophers of science and theoretical physicists have a professional duty to be interested in the interpretation of quantum mechanics—the latest, best established theory of physical processes. Even some members of the educated public, going above and beyond the call of duty, are interested. Unfortunately, philosophers and physicists have mostly—giants like Bohr, Born, Einstein, et al. surely excepted—written on the subject in needlessly abstruse fashion, in arcane symbolism and terminology, and thus prevented each other, much less the educated public, from understanding and resolving the very real problems involved. Further—again with notable, though different, exceptions—the physicists who have written on quantum theory have been innocent of philosophic competence, the philosophers devoid of ability in theoretical physics. For the above reasons, among others, much of the literature on quantum theory is either unreadable or irrelevant. Some is both. I hope that the present work is neither, to any significant degree, and that it will help bridge current communication gaps between theoretical physicists and philosophers of physics, and between both groups and the educated public.

To properly thank all those who helped bring the present work to fruition would require a separate volume, nearly as long as this one. I must, however, single out a few individuals, so that they may share whatever praise or blame attaches to this book. I take this opportunity to express my gratitude to the two readers of the Johns Hopkins doctoral dissertation on which this manuscript is based: Professors Peter Achinstein (Philosophy) and George Owen (Physics). Without their criticism, advice, and guidance, this work would now be far different, and definitely inferior. I am also indebted to the two anonymous readers for the University of Chicago Press, whose criticisms are responsible for significant improvements in both the style and content of the manuscript. Finally, I must mention my great debt to Max Born and Alfred Landé, the two precursors of the interpretation herein defended. Many of the arguments of this essay are, so far as I know, original, and my presentation, as a whole, seems to me far more complete and convincing than any previous defense of an indeterministic particle interpretation. Nevertheless, the fact remains that the basic conclusions of the interpretation, and the key insights needed to apprehend the truth of these conclusions, are due to Born and Landé.

1

The Copenhagen
Interpretation of
Quantum Mechanics

There is considerable confusion in the literature as to just what is the Copenhagen interpretation of quantum mechanics. One reads that the Copenhagen interpretation is a dualist interpretation in which subatomic entities are both waves and particles (Hanson 1967, pp. 41–49; Hanson 1963); that these entities are neither waves nor particles but somethings-we-know-not-what which exhibit wave-like properties in certain experimental arrangements and particle-like properties in other arrangements; or that they are particles obeying irreducibly statistical laws (Born 1964, 1956). This variety is surprising since all parties concerned seem to be attributing the Copenhagen interpretation to Bohr and Heisenberg.

In this chapter and the next I will give a critical exposition of the interpretation of Bohr and Heisenberg and, concurrently, show that the differences between their interpretation and later expositions, including many recent textbook discussions, are due to ambiguities in the key terms 'wave' and 'particle.' For well-known historical reasons the term 'Copenhagen interpretation' ought to be reserved for the views of Bohr and Heisenberg, but nothing hangs on the labeling so long as we distinguish among the various possibilities.

It is well to remind ourselves at the outset just what an "interpretation" of quantum mechanics consists of and why the problems of interpretation which arise in quantum mechanics appear more intractable than those arising in other physical theories. Some

1

philosophers of science claim that all terms used in physical theories are completely dependent for their meaning on a particular theory and hence in order to understand the meaning of these terms we must understand the theory in which the terms function. Other philosophers claim that only so-called "theoretical" terms are dependent for their meaning on the theories in which they function while other terms, "observation" terms, are independent of any theory and mean the same in all theories. If either of these views of theories and theoretical terms were correct, we would face formidable if not insurmountable difficulties of interpretation with regard to any and all theories.

Fortunately, neither view is tenable, and I show this untenability in chapter 3. For present purposes it is enough to note that quantum mechanics is a quite atypical theory in certain respects. Any difficulties involved in the interpretation of various classical theories—mechanics, electromagnetics, kinetic theory, thermodynamics, and so on—were soon overcome, while difficulties of interpreting quantum mechanics persist after several decades and persist undiminished. This difference is attributable to two factors, one conceptual and one empirical. We are able to understand the meaning of terms and the sentences, equations, etc. in which they function in classical theories because, typically, these terms are not introduced and used as uninterpreted or even "partially" interpreted terms. Rather, such terms are either explicitly defined —necessary and sufficient conditions are specified for their correct usage—or, more typically, some of the sentences, equations, etc. in which they function amount to important criteria for correct usage of the terms though not necessary or sufficient conditions. Nor are all these terms completely dependent for their meaning on the usages they have in any one theory. Quite the contrary, many terms common to two or more theories can clearly be said to have the same meaning in all these theories. This favorable conceptual situation with regard to classical physics is related to the empirical situation which, on the whole, was also a favorable one.

Great descriptive, predictive, and explanatory success was achieved in diverse areas of classical physics by using theories embodying one or the other of two kinds of models, particle models and wave models. The terms, formulas, sentences, etc. of classical physics typically presented no formidable problems of interpretation because their meaning could be explicated via some version or other of one of these two types of models. Such explication sometimes involved the literal attribution of certain "particle" properties to the referents of certain "theoretical" terms (e.g., in kinetic theory) and sometimes involved readily understandable analogues of familiar wave properties (e.g., in electromagnetics and physical optics). These (related) favorable situations do not prevail and have never prevailed in quantum theory.

Quantum theory evolved through attempts to present unified accounts of various specific physical phenomena, accounts whose very possibility was called into question by disconcerting theoretical and experimental discoveries in the early part of this century. Electromagnetic radiation, including of course light, was found to behave in ways seemingly only explicable via a particle model whereas formerly it had always behaved in wave-like fashion. Conversely, subatomic material particles which had, as their very name suggests, previously been found to behave like particles were found to also display wave-like behavior. Accordingly, in a superficial way we can describe the problems of interpreting quantum mechanics by saying that no interpretation of specific phenomena in terms of a single model seems satisfactory. If this apparent difficulty proves insurmountable, then it would seem that terms common to classical and quantum theory cannot have the same meaning in the two theories. But in that case what do the terms mean in quantum mechanics? Moreover, what is the relation between the two theories; how can they be compared if they have no common terms with theory-invariant meaning?

A word is in order here on the terms 'wave' and 'particle.' Notice that what counts as a wave or a particle, or even as a particle rather than a nonparticle, is a contextual matter. We should not suppose that a given physical object or process is, in and of itself, a particle or a wave, a particle process or a wave process. For example, it makes no clear sense to ask whether a baseball is or is not a particle, unless this is asked in the context of a given problem, calculation, measurement, analysis, etc. For some purposes a baseball is a particle; for other purposes it is a collection of particles; for still other purposes it is neither particle nor collection of particles but just (roughly) a spherical solid. For many, perhaps most, purposes an electron (or proton or neutron, etc.) is a particle; for most purposes a gas molecule is a particle even though in general such a molecule contains more than one atom which in turn consists of a nucleus and one or more electrons, the nucleus in general consisting of several nucleons—protons or neutrons.

The point here is that spatially extended objects can count as particles even if the objects consist of collections of smaller subsystems, *provided* the dimensions and hence the internal structure of these extended objects are negligible compared to the other dimensions of the problem at hand. With this in mind we can understand why baseballs count as particles for ballistic calculations, why planets count as particles in celestial mechanics, why stars or even galaxies count as particles for certain cosmological calculations. Thus if we claim that a particle model satisfactorily explains a certain phenomenon, we are not committed to the absurd claim that the items identified as particles are immutable and/or actually (as small as) geometric points, occupying no extended volume of space; we are only claiming that, for certain specific purposes, certain important properties of these items can be regarded as located at such a geometric point. These facts are so obvious I would blush to mention them were it not the case that some philosophers and even physicists have managed to forget or even deny them (Schrödinger 1935; Margenau 1950).

Corresponding remarks apply to the term 'wave.' In general a wave is any sort of disturbance—any noticeable or at least specifiable periodic variation in some physical quantity—which is transmitted in a continuous medium. This medium may or may not be a material one; the periodic variations may be spatial oscillations of a particle or particles, they may be oscillations in the strength of a field or fields at spatial points located in a vacuum, or they may be of some third kind. Any physical process may be regarded as wave-like provided only that it exhibits the requisite properties of superposition and periodicity. These features in themselves neither forbid nor require further descriptions of the physical items involved.

With the above preliminaries in mind, let us review some of the highlights of the evolution of early quantum theory. The quantization of action or, equivalently, the fact that the energy values associated with certain kinds of physical systems are discrete rather than continuous was discovered in 1900 by Planck in the study of blackbody radiation. Bohr used this idea of Planck's in his theory of atomic structure which explained the existence of line spectra. Bohr assumed that the atomic system emitting the radiation can exist only in discrete energy states—states intermediate to these "eigenstates" are impossible. Radiation is then emitted as a result of transitions from one eigenstate to another, the radiation carrying away the energy lost by the atom, with frequency proportional to this energy, Planck's constant (h) being the proportionality factor.

The most surprising aspect of these discoveries was the fact that the energy (E) of the electromagnetic radiation involved was shown to be proportional to the frequency (ν) of the radiation, $E = h\nu$. Previously, it was supposed that these quantities (E,ν) were independent. Equally surprising was the relation between the momentum (p) of electromagnetic radiation and its wavelength (λ), $p = h/\lambda$. This latter relation was discovered in the relativistic analysis of the Compton effect wherein radiation is scattered by electrons, and transfers part, but not all, of its momentum and energy. De Broglie later suggested that the relation $p = h/\lambda$ holds for material particles, and this suggestion was confirmed by the electron diffraction experiments of Thomson and of Davisson and Germer. But if we are led to associate waves with material particles, then it is plausible to suppose that there should be a wave equation governing the behavior of the waves.

Furthermore, a striking *formal* analogy between classical mechanics and geometrical optics had been known for nearly a century. But geometrical optics, a ray theory, is known to be only an approximation of physical optics, a wave theory; that is, in the limit of infinitely small wavelengths the wave equations of physical optics approach the ray equations of geometrical optics, the rays in general being perpendicular to the wave fronts or surfaces of equal phase. In view of these facts it is plausible to suppose that, analogously, classical mechanics might be only an approximation to, a limiting case of, a more general mechanical theory, a theory which, like physical optics, is basically a wave theory. It was just such a theory which was developed by Schrödinger and embodied in the two basic differential equations of quantum mechanics which bear his name.[1]

But what are we to make of these equations and the wave-like functions which are their solutions? How seriously are we to take the cited analogy? Should we continue to regard it as a purely formal or mathematical analogy, or should we interpret the wave equations and their solutions as representing the oscillations of some actual physical items, whether particle position(s), charge or mass density values, field strengths, or other items? Or should we, on the other hand, interpret the wave functions or quantities related to them in terms of probability distributions of particle properties and not in terms of actual physical oscillations? Or should we reject both these alternatives in favor of some third interpretation? It is questions such as these which interpretations of quantum mechanics are designed to answer. These questions are closely related to another question concerning the significance of the Heisenberg uncertainty relations, perhaps the most important question of all: are the uncertainty relations due to the "dualist" (simultaneous wave plus particle) character of matter and light or is the converse true or both or neither? In any case are the uncertainty relations explained by or explanatory of, or both, the disturbance of subatomic systems by measurements? Let us see how Bohr and Heisenberg answer the questions set out in this paragraph. For various historical reasons we may safely take Bohr's 1927 Solvay Congress paper (Bohr 1961) and Heisenberg's 1929 Chicago lectures (Heisenberg 1930) as early and authoritative statements of the position in question.

Bohr begins the body of his paper by stating his conclusions: that quantum theory demands a "limitation" of classical ideas when applied to atomic phenomena, that this limitation is due to the essential "discontinuity" or "individuality" of atomic processes and that the limitation appears most peculiar because these classical ideas are the only possible ones available for describing experimental results (Bohr 1961, p. 53). Bohr holds that quantum theory implies that any observation of atomic phenomena will involve a nonnegligible interaction with the agency of observation. He claims, furthermore, the distinction between the system to be observed and the agency of observation is ultimately arbitrary. He goes on to say that specification of the "state" of a physical system requires that the system be isolated, that all external disturbances be eliminated. In that case, however, no observations are possible and, in particular, no determination of the spatio-temporal properties of the system is possible. On the other hand, if observational interactions are allowed, then no "unambiguous" definition of the state of the system is possible and no causal description of its spatio-temporal development is possible. Hence spatio-temporal descriptions on the one hand and causal description on the other are said to be "complementary but exclusive" features. The appropriateness of our usual classical, causal, *and* space-time description is due to the small value of Planck's constant of action (h) compared to the actions involved in macroscopic phenomena.

As the actual starting point of his argument, Bohr cites the apparent dual aspects of both matter and light as indications of the complementary but exclusive character of causal and space-time descriptions. His idea here is that an exact space-time description in terms of waves is possible so long as radiation does not interact with matter; when such interaction does take place, a quite different description, in terms of light quanta or photons, is necessary. But it turns out that this new "causal" description is incompatible with an exact space-time description. We can only correlate statistically the number of photons arriving at a point in space with certain properties of the electromagnetic wave at this point—we cannot trace the motion of individual photons from the light source to the point of detection. Conversely, in the case of material particles, certain phenomena such as the "interference" involved in scattering off crystal lattices requires the wave theory superposition principle. Here again the idea is that an exact description of the motion of, say, electrons from source to detector seems to require a wave model; but the causal interactions involved at the detector seem to require a particle model. So, both with regard to matter and light, the causal, particle description is complementary to and (strictly speaking) exclusive of the spatio-temporal wave description.

Following this introduction of the notion of complementarity, Bohr attempts to show that this notion is implicit in "the most elementary concepts employed in interpreting experience." He begins (Bohr 1961, p. 57) by citing the formulas underlying both the theory of light quanta and the wave theory of matter

$$E/\nu = p\lambda = h, \tag{1}$$

wherein concepts associated with particles having definite space-time coordinates—the concepts of energy (E) and momentum (p)—are connected via Planck's constant (h) to the wave concepts frequency (ν) and wavelength (λ) which, strictly speaking, "refer to a plane harmonic wave train of unlimited extent in space and time." This last point is of crucial importance. We cannot represent a quasi-periodic process, i.e., one localized in space or time, by a genuinely periodic function—for such a function has nonzero values outside the locale of interest. Or, to put it another way, the process does not really have a wavelength or frequency since it does not repeat itself outside the localized region. But we can think of this localized process as having an infinite wavelength and period (inverse frequency) and represent its function as the sum of a continuous series of genuinely periodic functions corresponding to genuinely periodic processes.

Because of the principle of superposition of amplitudes, one of the defining properties of wave processes, we can choose a set of elementary (genuinely periodic) waves whose amplitudes in effect cancel each other outside the localized area of interest. But now each member of this representing set has a (slightly) different wavelength and frequency. It is a theorem of mathematics, a corollary of the Fourier integral theorem, that this "spread" in frequency and wavelength is inversely proportional to the size of the localized region. Hence the smaller the localized region the greater the spread in frequency and wavelength; the larger the region the smaller the spread; and as the size of the region increases without limit the spread approaches zero—for a completely unlocalized wave process and only for such, a single wavelength and frequency are defined.

Now it is easy to see how equation (1) leads directly to some of Heisenberg's uncertainty relations. For if we try to associate a unique, exact frequency and wavelength with an individual physical system, then this system must be assumed completely unlocalized—exact specification of ν and λ and hence of the energy and momentum of the system implies that the system extends over all time and space. Conversely, exact specification of the position of the system—localization to a geometric point at an instant—implies an unlimited spread in frequency and wavelength and hence in energy and momentum. Thus, qualitatively, we see that the uncertainties in spatial position and in time are inversely proportional, respectively, to the uncertainties in momentum and energy. Quantitatively it turns out that

$$\Delta x \cdot \Delta p_x \geqq h,$$
$$\Delta E \cdot \Delta t \geqq h, \tag{2}$$

where x stands for any of the three spatial coordinates. So, according to Bohr, it is equation (1) which forces the limitation of our classical concepts, as expressed by equation (2).

Bohr follows his derivation of equation (2) with the remark (Bohr 1961, p. 60) that the limitation of classical concepts expressed by equation (2) is connected with the limited validity of classical mechanics. Just as geometrical optics is a limiting case of physical optics, so classical mechanics is a limiting case of wave mechanics, and only in this limit can energy and momentum be associated with individuals having exact space-time coordinates. In general, the "definitions" of these concepts (E,p) is given entirely by the conservation laws. However, if we replace the idea of particles with exact space-time coordinates by that of "unsharply defined individuals within finite space-time regions," then we can, Bohr alleges, retain a sort of approximate space-time model of processes involving energy and momentum transfer, such as the photoelectric and Compton effects, as well as collisions between material particles.

Finally, Bohr notes that the discontinuous change of energy and momentum which accompanies some measurements would not, in itself, prevent the assignment of exact space-time coordinates both before and after the process of interaction involved in the measurement. Rather, the impossibility of such assignment is due to the fact cited above—the inverse relation between spread in frequency and wavelength and size of the region of localization of the wave process which represents the energy and momentum carriers. Thus we could avoid "disturbing" the system under observation by letting it interact with either radiation or material particles, provided the interacting item had small enough momentum or, equivalently, large enough wavelength associated with it. But then the position of the observed system would be uncertain within this large wavelength.

So Bohr is claiming that the uncertainty relations are not just a consequence of the disturbance involved in atomic measurement and that such disturbances can be made increasingly smaller. Thus the slogan "observation disturbs the observed system in an unpredictable way" is misleading but correct, if we understand "observation" to mean *successful, accurate* observation of one of a pair of conjugate variables and understand disturbance to apply to the other member of the pair.

9

In view of the above, the essential point of Bohr's interpretation of the Schrödinger wave equations and their solutions is seen to be that "radiation in free space as well as isolated material particles are abstractions, their properties on the quantum theory being definable and observable only through their interaction with other systems" (Bohr 1961, pp. 56–57). For heuristic purposes, it is convenient to note how this point applies to Schrödinger's own interpretation of the theory.[2] The characteristic (linearly independent or orthogonal) solutions of the Schrödinger equation furnish exact representations of the stationary states, i.e., energy states, of an atomic system. Schrödinger associates with the solution of the wave equation a continuous distribution of charge and current and, presumably, mass. In the case of the characteristic solutions, this distribution represents the electrostatic and magnetic properties of the atom in this stationary state. The superposition of two or more characteristic solutions corresponds to a continuous distribution of vibrating electric charge, which would emit radiation just as expected by classical electrodynamics.

Thus we can think of the general case, the dynamic one, as represented by oscillations in a continuous medium, while the special stationary case corresponds to a standing wave pattern within this medium. Furthermore, Schrödinger claims that the discontinuous exchange of energy between atoms, one of the sources of the troublesome "dualist" relations in equation (1), can be treated as a simple resonance phenomenon. That is, consider two atoms, I and II, with characteristic energy spectra $E_I(n)$ and $E_{II}(i)$ and at least one common frequency $\nu_I(nm) = \nu_{II}(ik)$

$$h\nu_I(nm) = E_I(n) - E_I(m),$$
$$h\nu_{II}(ik) = E_{II}(i) - E_{II}(k),$$

so that $E_I(n) - E_I(m) \doteq E_{II}(i) - E_{II}(k)$. According to Schrödinger, the "resonance" explanation of the energy exchange between I and II is that the two atoms are continuously exchanging energy, with one going through an energy (vs. time) maximum while the other goes through a minimum, and vice versa. Of course, the reason for the energy exchange is that the two atoms are considered coupled oscillators, each providing an external driving frequency which is in certain phase relations to the natural frequency of the driven oscillator. In such a coupled system the amplitude of each oscillator will go through sharp periodic maxima while the other goes through its minima, provided the driving and driven frequencies are equal.

Bohr (1961, pp. 75–79) and Heisenberg (1930, pp. 47–54, 157–83) object to this interpretation of Schrödinger's on the ground that both in the case of stationary energy states and in the "resonance" phenomena involved in energy exchange between atoms we are dealing with "isolated" systems whose spatio-temporal behavior is accordingly unknown to us. Once again the argument is that, strictly speaking, the wave function associated with an exact atomic energy state is completely unlocalized; hence it cannot be said to represent the actual structure of this (very small) atom. Conversely, if the wave function is localized via representation by a set of functions, then it no longer represents exactly a single energy—a stationary state. Similarly, in the "resonance" exchange between atoms, both the spatial location of the resonances and their temporal development are completely undetermined if the resonances are representative, via the frequencies involved, of exact energy states. We cannot trace the energy exchange via spatio-temporal developments in this case. Conversely, if we localize the resonances and trace their temporal development, then they no longer represent exact energy states.

So Bohr and Heisenberg would reject any "realist" wave interpretation, such as Schrödinger's, in which the space-time coordinates of some actually oscillating items could be specified exactly.

It is important to realize that these two men reject just as strongly and for the same reasons any particle interpretation, even a "statistical" one. They take "particles" to be individual carriers of energy and momentum. Accordingly, they would consider it misleading to say that (the square of) the amplitude of the wave function represents the probability of finding an individual particle at the location corresponding to this amplitude or, in the case of many particles, to the relative population at this location—relative to the average over all space. That is, if we associate an exact energy and momentum with an individual particle, then, according to the arguments discussed above, its position is completely undetermined; there is no more probability of finding it in one place than in any other.

Similar remarks apply to collections of (independent) particles. In order to be acceptable to Bohr and Heisenberg, the statistical "particle" interpretation would have to be restated in one of two forms. It could be stated that the carriers of energy and momentum are localized within a certain region of space, at the price of a corresponding spread in the value of the energy and momentum carrier. Within this localized region, nothing could be said about where or even what these "carriers" are. Conversely, we could represent the passage of momentum and energy through space by a wave function with exactly specified wavelength and energy. Then the energy density of the wave process at a spatio-temporal location is proportional to (the square of) the amplitude, $|\psi|$, of the wave. But the probability of momentum-energy transfer at this location is also proportional to the energy density. Hence, if we assume the transfer to take place via individual carriers, the (relative) number of these carriers is proportional to $|\psi|$.[2] But the motion of these carriers along the wave cannot be pictured spatio-temporally as can that of the wave itself; they are only correlated with it.[3] It is only in the classical limit of high momentum and energy that the carriers can be assumed to move perpendicular to the wave-front. In neither of these two reformulations would we be using the word 'particle,' referring to the carriers of energy and momentum, in a strictly correct manner according to Bohr and Heisenberg. For they take a particle to be an item with exactly specifiable simultaneous momentum and position or, equivalently, energy and time.

So we can see that, far from interpreting atomic and subatomic systems as both waves and particles, Bohr and Heisenberg refuse to interpret them as either. We have seen that they reject Schrödinger's wave interpretation and any other interpretation which takes the wave equations and their solutions as representing the actual space-time oscillations of some physical quantity. Furthermore they would, as we have seen, reject statistical "particle" interpretations presented in various textbooks unless these were reformulated to indicate the loose nature of the application of the word 'particle.' Having reviewed their rejections, and the reasons for these rejections, of both wave and particle interpretations, we can now see the point of Bohr's remark, quoted above, that both radiation (i.e., waves) and particles are "abstractions" whose properties are specifiable only through their interactions with other systems. The systems represented by the solutions of the wave equations—in general by an appropriate set of waves—are physical systems necessarily without any well-defined wave or particle properties. They lack such well-defined properties precisely because they are represented by such a set of waves, as we saw above. The meaning of the last part of Bohr's claim—that the properties of these systems are only specifiable through interactions with other systems—is crucial for an understanding of the Copenhagen interpretation.

The meaning of the passage in question can be brought out by comparing it with typical textbook formulations of one of the axioms of quantum mechanics. Accurately paraphrased (cf. Mandl 1957, p. 66), this axiom tells us that if a certain kind of measurement—one attempting to find the value of a particular variable such as, say, momentum—is carried out on a physical system represented by a given wave function—in general by an appropriate *set* of waves—then the probability of finding a value of the momentum within a given interval depends on the number and relative weighting of the eigenstates of momentum within this interval. That is, in general the wave function representing the system to be measured will not be an eigenfunction corresponding to a momentum eigenstate but will be a linear combination of such eigenfunctions, the coefficients of the linear expansion representing weighting factors for the various eigenstates. Whenever a particular measurement is performed, one definite momentum value is found, but Bohr wants us to notice two crucial points about this measured value.

First, no exact predictions about all future behavior of the system can be inferred from the measured value, because in determining an exact momentum value we have—or rather the measuring apparatus has—disturbed the conjugate position variable in an unpredictable way. Corresponding remarks apply to other pairs of conjugate variables. Second, and more important at this juncture, Bohr tells us that we can not infer that the particular measured value was the value that the system had before measurement. For if we repeat the measurement—measure other systems also represented by the same wave function as our original system, and hence indistinguishable from it so far as the variables being measured are concerned, or remeasure the original system after somehow bringing it back into the state represented by the original wave function —then we will not in general get the same results as in the original measurement. What we will get is a spectrum of values whose average is uniquely determined by the wave function representing the original state and by the representation of the variable being measured—i.e., by the mathematical operator corresponding to this variable. It is only in certain special cases that repeated measurements will all yield the same numerical result. For instance, if the system was originally in one of its eigenstates, then repeated measurements will all yield the same numerical result, namely, the eigenvalue corresponding to this eigenstate.[4] To put it another way, the wave function representing the system will in this special case consist only of a single term corresponding to the one eigenstate; the coefficients or weighting factors for the other eigenstates will all be zero for this particular expansion of the wave function. Hence the probabilities of finding the eigenvalues corresponding to these eigenstates will be zero.

Even from this brief sketch of the quantum theory of measurement, we can see the point Bohr is driving at. In general, physical systems will be represented by wave functions such that the properties of these systems possess a certain indefiniteness. Measurement of such systems will sometimes yield one result, sometimes another. The only definite properties assignable to these systems are those which are the outcome of the measuring process and the disturbance it involves. Furthermore, even in the case of systems which are in eigenstates before measurement, the only way to know this fact is to have previously performed repeated measurements and to have obtained always the same results; otherwise it would not be known if the system should be represented by a single term (eigenstate) wave function or by a more general linear combination of such functions.

So, Bohr concludes, in all cases to attribute a definite, exact property—value of a variable—to a system is an elliptical way of saying that the system responds or has responded in a certain way to measurements of the variable corresponding to this property. In the general case we can only attribute average properties or "expectation" values to systems; in the special case of eigenvalues we are elliptically attributing to the system the genetic property of having yielded the same results upon successive measurement and hence the future capacity to do so. Conversely, Bohr claims, if we want to attribute a property to a system independently of its interaction with other systems we can do so only *approximately*, specifying the value of one conjugate variable to good accuracy only at the expense of a corresponding error in the other variable of the conjugate pair.

Unfortunately, there is no neat way of characterizing this Copenhagen interpretation that we have partially explicated. The interpretation is to a large extent negative in character. It denies what other interpretations assert. It denies that the wave equation solutions represent the actual oscillations of any physical quantity. It denies that they represent, implicitly, probability distributions of particle properties, where 'particle' is used as in classical mechanics. Insofar as it has a positive content, it might be called, for lack of a better name, a "relational" interpretation since it claims the only objectively specifiable properties of physical systems are those they exhibit with respect to various kinds of interactions with other systems—notably including measurement interactions.

It would be a grave mistake to attempt to apply various standard philosophical labels to the Copenhagen interpretation. It is clear enough that in a sense it is a nonrealist or even antirealist interpretation. It not only denies that physical systems can be described exactly in terms of either a particle or a wave model—it claims that the only exact description forbids the attribution of intrinsic properties and allows only the attribution of relational ones. But it might be quite misleading to call the Copenhagen interpretation a nonrealist or antirealist one: Bohr and Heisenberg do not deny that the physical systems in question *exist* independently of measurement or other interaction, they merely deny that any classical model exactly describes their intrinsic properties. (What then do these classical models describe? Bohr calls them "abstractions" to indicate they are correct only in limiting cases which never quite apply to any actual physical systems.)

It would be even more misleading to label the Copenhagen interpretation an "instrumentalist" one. Of course, this label would bring out some features of the interpretation well worth emphasizing. For instance, that the interpretation denies that any specific realist interpretation—beyond the minimal one that quantum mechanical systems *exist* independently of measurement or other interaction—is necessarily false; further, it reminds us that interactions with measuring instruments play a crucial role in the interpretation. Still, 'instrumentalism' has connotations such that the use of the term here would be misleading and should be avoided. First, 'instrumentalism' has a technical usage in the philosophy of science, referring to the view that theories are instruments or calculational devices as opposed to truth claims about the items corresponding to (some of) the theoretical terms. Second, 'instrumentalism' is sometimes used synonymously with 'operationalism,' the view that the meaning of all scientific terms is reducible to the results of measurement operations. In the early works under consideration, Bohr and Heisenberg do not assert nor are they committed to defending "instrumentalism" in either of the two senses cited above. Accordingly, even if they are sympathetic to one or both of these views, we cannot attack their interpretation by attacking either form of instrumentalism since neither form is presupposed by any of their actual analyses.

One final point remains to be made, concerning the significance of the "correspondence principle." Briefly, the principle is that for large quantum numbers—large values of energy and momentum or of "action" as measured in units of h—the quantum theory will approach classical theory as an asymptotic limit. Thus for macroscopic phenomena quantum effects can safely be neglected since the difference between the predictions of quantum theory and classical theory will be much smaller than purely experimental errors. This, according to Bohr, is why experimental results can be, as they must be, described in classical terms.

Now, some commentators (Putnam 1962b; Nagel 1961) have insisted that Bohr and Heisenberg hold that quantum effects disappear entirely on the macroscopic scale. If this just means that such effects are undetectable, the interpretative claim is innocent enough. If, on the other hand, the claim is that Bohr and Heisenberg think that, theoretically, quantum effects are not present on the macroscopic scale, either because they have somehow "cancelled out" or otherwise attained zero magnitude, then the interpretation flies in the face of clear statements to the contrary by Bohr and Heisenberg. For example, Bohr says:

In fact, the connection in question between quantum and classical theory means that in the limit of large quantum numbers where the relative difference between adjacent stationary states vanishes asymptotically, mechanical pictures of electronic motion may be rationally utilized. It must be emphasized, however, that this connection cannot be regarded as a gradual transition towards classical theory in the sense that the quantum postulate would lose its significance for high quantum numbers. On the contrary, the conclusions obtained from the correspondence principle with the aid of classical pictures depend just on the assumptions that the conception of stationary states and of individual transition processes are maintained in the limit (Bohr 1961, p. 85).

This quotation, because of the context in which it occurs and because it mentions electrons, might be regarded as intended to apply only to atomic and subatomic systems, but other passages, free from such ambiguities, are readily available. Thus, Bohr says:

In connection with the measurement of the position of a particle, one might, for example ask whether the momentum transmitted by the scattering could not be determined by means of the conservation theorem from a measurement of the change of momentum of the microscope—including light source and photographic plate—during the process of observation. A closer investigation shows, however, that such a measurement is impossible, if at the same time one wants to know the position of the microscope with sufficient accuracy. In fact it follows from . . . the wave theory of matter that the position of the center of gravity of a body and its total momentum can only be defined within the limits . . . given by [the uncertainty relations] (Bohr 1961, p. 67).

Similarly, Heisenberg reminds us that in determining the position of a particle by observing it through a microscope

we might seek to determine the path [of the particle] by making the microscope movable and measuring the recoil it receives from the light quantum [used to determine the particle's position]. But this does not circumvent the uncertainty relation for it immediately raises the question of the position of the microscope, and its position and momentum will also be found to be subject to [the uncertainty relations] (Heisenberg 1930, p. 22).

Thus both Bohr and Heisenberg clearly and explicitly state that quantum effects exist for large pieces of hardware such as microscopes as well as for atomic and subatomic systems. With this point we may conclude our exposition of the Copenhagen interpretation. Exactly what, if anything, is wrong with this interpretation of Bohr and Heisenberg?

2
Objections to the Copenhagen Interpretation

Misguided Criticisms of the Interpretation

While it is important to appreciate the unsatisfactory character of the Copenhagen interpretation, it is equally important to realize that many popular criticisms of the interpretation are unfair, irrelevant, or otherwise misguided.

The first misguided objection concerns the allegedly "subjective" character of the Copenhagen interpretation. Consider the following quotation from Bunge:[1]

Yet in recent years the tide has begun to turn, not only in philosophy but also in physics. Within the so-called Copenhagen interpretation itself, certain objectivist trends are appearing. Thus while Bohr . . . had always insisted that the epistemological lesson of atomic physics was that the object could not be separated from the subject, Heisenberg . . . has recently admitted that "it does not matter whether the observer is an apparatus or a human being," and that "the introduction of the observer must not be misunderstood to imply that some kind of subjective features are to be brought into the description of nature." The days in which it was believed that "ordinary (i.e. macroscopic) phenomena are in a way engendered by repeated observations" seems to be over—well, almost over. The physicist of the latest generation is operationalist all right, but usually he does not know, and refuses to believe, that the original Copenhagen doctrine—which he thinks he supports—was squarely subjectivist, i.e. nonphysical (p. 4).

This single short passage contains no fewer than five major exegetical mistakes. Apart from its convenient compactness, this remarkable performance is not at all untypical of the writings of a

18

certain school of Copenhagen critics, represented by Bunge and Popper. Let us consider each of these mistakes in turn.

First, there is the false implication that Heisenberg's "admission" represents a recent change of opinion on the matter of the necessity of human observers. Bunge cites a 1958 book by Heisenberg but fails to tell us that the "admission" in question was made by Heisenberg in 1929:

The observing system need not always be a human being; it may also be an inanimate apparatus, such as a photographic plate (p. 58).

Second, we have the false implication that Bohr's position all along has been contrary to this "recent admission" of Heisenberg's. But, in the 1927 paper we have been examining, Bohr tells us:

In tracing observations back to our sensations, once more regard has to be taken to the quantum postulate in connection with the perception of the agency of observation, be it through its direct action on the eye or by means of suitable auxiliaries such as photographic plates, Wilson clouds, etc. (1961, p. 67).

Third, we have Bunge's treatment of Bohr's claim that the distinction between observed and observing system is ultimately arbitrary. Bunge misleadingly translates this as the claim that "the object could not be separated from the subject"—thus begging the exegetical question by bringing in connotations of consciousness and mind-dependence via the phrase 'the subject'—and implies that the original claim somehow supports the claim that conscious observers are necessarily involved in measurement interactions. Just the opposite conclusion should be drawn. If the boundary between the observed and observing system is arbitrary, then we can draw this boundary by isolating a single atom or part thereof as the observed system so that the observing system is then just the rest of the experimental apparatus involved and has nothing to do with the experimenter. Should all experimenters—all conscious beings— cease to exist, the boundary between observed and observing systems would still be equally arbitrary, and the interaction between the two systems would still disturb the observed system in an unpredictable way, according to Bohr and Heisenberg.

Fourth, we can now vaguely see what sort of view Bunge is attributing to the Copenhagen interpretation.[2] In the last quoted sentence he equates "subjectivist" with "nonphysical" and it is clear from other passages in the same paper that "nonphysical" is to be equated with "mental" or "mind-dependent." In the final sentence of his paper Bunge implies that what the critics of the Copenhagen interpretation have in common is the belief "that there is an autonomous external world" and, presumably, that the view they are united in opposing is one which denies this autonomous existence. But we have seen that Bohr and Heisenberg do not deny and are not committed to denying that subatomic particles—or any other physical systems—exist independently of human conscious-ness and human actions. What they deny is that any literal attribu-tion of various (e.g., wave or particle) classical properties to these systems could possibly be correct, and this denial is quite different —much weaker, more sensible and defensible—than the other one.

Fifth, Bunge quotes out of context Heisenberg's remark about macroscopic phenomena being engendered by repeated observa-tions, and, in his annotated list of references, says Bohr quotes this remark with approval. Heisenberg and Bohr do both hold that quantum effects apply to macroscopic phenomena, but this point is not the one at issue. The question is whether all phenomena and macroscopic phenomena in particular are created by observation. It is clear from the context of the original remark and the context of Bohr's approving citation of it (Bohr 1961, pp. 67–68) that Heisenberg is using the familiar fact that actual measurements yield a spectrum of values, the dispersion being due to random errors if nothing else, as a *classical analogue* of the quantum mechanical feature that atomic and subatomic entities have no intrinsic (wave or particle, etc.) properties independent of measurement (or other) interactions. That is, Heisenberg is only illustrating the quantum mechanical point by noting a certain aspect of the classical macro-scopic situation, using the latter as a heuristic device. Just as we only know the quantitative properties of macroscopic objects by measuring a spectrum of values and associating "the value" with the most probable value of the spectrum, so too in quantum me-chanics we can assign no intrinsic values independent of measure-ment interactions. Of course the analogy is a very weak one, as Bohr is quick to point out. In classical physics it was assumed that in the limiting case of infinitely repeated measurements the dis-persion in the results could be made to vanish; i.e., in principle each measurement increased the accuracy of our knowledge of "the value" of the property being measured. In quantum mechanics this assumption is held to be false. On the contrary, each successive measurement disturbs the measured system in an unpredictable way. So when Bohr and Heisenberg speak of any phenomena, macroscopic or atomic, being created by observation they do not mean that the measured system does not exist and undergo transformations and interactions independently of observation. What they mean is that to attribute properties to these systems is either to be, at best, only approximately correct or to attribute, in an elliptical way, relational rather than intrinsic properties to the systems. The latter claim is controversial enough but we must not confuse it with the quite implausible former claim, one that Bohr and Heisenberg neither make nor commit themselves to.

The straw man attacked by Bunge is set up somewhat more clearly by Popper, who characterizes the Copenhagen interpretation as the view that

"objective reality has evaporated," and that quantum mechanics does not represent particles, but rather our knowledge, our observations, or our consciousness of particles (Bunge 1967, p. 7).

Popper further claims that this view, as *he* characterizes it above, is almost universally accepted and that his work is an attempt to "exorcise the ghost called 'consciousness' or 'the observer' from quantum mechanics" and to show that quantum mechanics is objective, not subjective. It is revealing that the quotation and paraphrase of Heisenberg which Popper uses in his characterization of the Copenhagen interpretation is from a popular essay written by Heisenberg in 1958. We are asked to believe that to be committed to the Copenhagen interpretation it is neither necessary nor sufficient to accept the technical analyses given by Bohr and Heisenberg in 1927–29; that, on the contrary, it is necessary and sufficient to accept the popular writings of Heisenberg in 1958, wherein he seems to assert things he explicitly denied thirty years earlier. It is just false that almost all physicists accept Heisenberg's Sunday supplement efforts when they accept the Copenhagen interpretation. They accept, essentially, the conclusions of the technical arguments presented by Bohr and Heisenberg in 1927–29.

This misleading label, 'subjective' or 'subjectivist,' that Bunge and Popper have tagged to the Copenhagen interpretation is related to another misguided criticism, one which claims that the interpretation derives from the general "positivist" and "antirealist" bias of

Bohr, Heisenberg, and (most) other physicists. In his introductory paper, Bunge sketches, quite accurately, the parallel development of physics and its philosophy in this century—a development in which most physicists from the late twenties to the present accepted both the Copenhagen interpretation and the two characteristic positivist tenets, verificationism and operationalism. Verificationism is the view that nonanalytic statements are meaningful if and only if some observational evidence could directly or indirectly falsify or, to some extent, confirm the statements. Operationalism is the view that certain measurement or observation operations specify necessary and sufficient conditions for the meaning of all scientific terms. But from the fact that Bohr and Heisenberg, among others, accepted these positivist doctrines does it follow that their interpretation of quantum mechanics is based on or presupposes these doctrines, as Bunge implies?

If this charge were true, it would be a damaging one. Verificationism sounds innocent enough, but this innocence is directly proportional to the vagueness with which I have stated the principle. In view of the demonstrated inadequacy of all precise formulations of the principle of verification which have been advanced (see Ayer 1952; Hempel 1950; Evans 1953), we must conclude that this view is dubious at best. Indeed, some of the original proponents of verificationism have been among its most acute critics. Similarly, operationalism has been effectively criticized in the literature and these criticisms need no rehashing here (see Lindsay 1937; Hempel 1954, 1966). However, the unattractiveness of these positivist tenets cannot count against the Copenhagen interpretation, since the actual arguments of Bohr and Heisenberg nowhere presuppose these tenets.

It is likely that this misunderstanding of the views of Bohr and Heisenberg is based on the many passages where they speak of the results expressed by the uncertainty relations as inherent in the nature of measurement or observation and on other passages where they speak of the meaninglessness of speculation about situations which are, in principle, unmeasurable. Both of these ways of speaking are unfortunate. First, consider the unobjectionable claims that measurement requires interaction with the system whose properties are being measured, that properties of isolated systems cannot be measured. From these facts nothing follows about the impossibility of exact simultaneous determination of all relevant kinematic and dynamical properties of a system or about the unpredictability of the disturbance due to the measuring apparatus. If the frequency and wavelength of radiation used in observations were completely independent of energy and momentum—if radiation of very short wavelength *and* very small momentum could be produced—then the disturbances involved in measurement could be made increasingly smaller and, equally important, any residual disturbances could be corrected for in an exact manner. But it is just because of the failure of this classical assumption about the independence of (ν,λ), on the one hand, and (E,p), on the other, that we are led to admit the existence of the uncertainty relations and the consequent unpredictability of the measuring disturbance. Hence it is false to say that these novel results are inherent in the nature of measurement or observation in the sense that whatever are our best-established theories about all physical processes, including those involving measurement interactions, these results will be implicit in such theories.

Of course, theories are used in interpreting and describing the physical interactions involved in measurements, and hence our present theory of measurement—which is just another way of saying our general physical theories as applied to measurement interactions—implicitly contains the results in question, the uncertainty relations. But to say that this means the results are "inherent in the nature of measurement" is a misleading way of making the above point. The results are inherent in our present theories, but theories may come and go, and have come and gone; hence our detailed theory of measurement will change and has changed. To say that the results are inherent in the *nature* of measurement is to lend our present theories a timeless quality which, considering the nature of theories, they could not possess. So, if interpreted literally, the cited passages about the nature of measurement are false. However, since Bohr and Heisenberg are quite clear and specific about the use of equation (1) in their arguments, about the fact that it is these equations which lead to acceptance of the uncertainty relations, the cited passages should be regarded as mere careless writing. Nowhere is the literally false claim about the nature of observation used in an actual argument.

Second, when they speak of the meaninglessness of speculation about unmeasurable properties it should by now be clear from the overall structure of their arguments that Bohr and Heisenberg are not relying on any dubious verificationist criterion of meaningfulness. In the cited passages on meaningless speculation they intend the "meaninglessness" to refer to the fact that certain combinations of simultaneous properties of physical systems cannot, strictly speaking, be specified. Again, the impossibility of this specification is built into the theory via relation (1). We might well be suspicious of this consequence and, accordingly, attempt to reinterpret these relations in a less straightforward way. At this juncture, however, the thing to notice is that the "meaninglessness" alluded to does not follow from the impossibility of measurement, but that rather the converse is true. Bohr and Heisenberg are claiming that whatever claims about physical systems cannot be coherently stated *using a quantum mechanical,* i.e., *wave functional, description* are meaningless. This is a suspect claim, but it is *not* a verificationist claim. The claim is that, because of the way physical systems are described, we cannot hope to measure (exactly) certain combinations of properties; this is just the opposite of the verificationist claim that whatever claims cannot (however indirectly) be observationally tested cannot represent meaningful assertions.

The best way to see the difference between the verificationist claim and the claim of Bohr and Heisenberg is to consider the specification of individual properties rather than combinations of properties. Recall that even classically, completely apart from quantum considerations, the exact wavelength of an oscillation cannot be measured. When we specify a plane wave by the theoretical description

$$\exp (i\mathbf{k} \cdot \mathbf{r}) = \exp (i2\pi \hat{k} \cdot \mathbf{r}/\lambda),$$

this wave, of length λ, extends over all space. As soon as we try to measure λ we restrict the wave to some finite region of space and hence introduce some dispersion in the wavelength. A verificationist would seem to be committed to saying that it is meaningless to talk about the (exact) wavelength of a wave because this property cannot be measured. Bohr and Heisenberg are quite willing to accept as meaningful, talk about exact wavelengths even though these items are unmeasurable. It is only talk about simultaneous wavelength and wavefront-position values that they regard as meaningless. Their criterion of meaningfulness is not what can be measured but, as we have now seen, what can be specified by a correct, i.e., wave functional, theoretical description.

The third and final misconstrual of the Copenhagen interpretation takes its "dualism" as referring to a "realist" interpretation involving both waves and particles. Incredibly enough, Hanson, a self-styled defender of the Copenhagen interpretation, appears to attribute to the interpretation the absurd view that subatomic entities can be both waves and particles. He tells us that

Newton's theory of fits—a theory which associated corpuscles of light with waves or "pressions" within the ether through which the corpuscles traveled—would have allowed light to have been simultaneously both undulatory and particulate. But in the nineteenth century the very ideas of particle and wave came to be fashioned in logical opposition to each other. . . . Had Newton's theory of fits prevailed, the choice between waves and particles might not have been made exclusive; we might have been conceptually comfortable with entities possessing both undulatory and granular characteristics. . . . phenomena which were incapable of being described exclusively in either purely wave or purely particle terms were discovered between 1887 and 1927. These phenomena required using a descriptive apparatus which was simultaneously appropriate to the wavelike and to the particulate properties of single objects (1967, pp. 41–49).

Two points should be noted here. First, Hanson has misinterpreted Newton's theory of fits. Only by identifying "light" simultaneously with the corpuscles and also with the oscillations of ether pressure, or density, and yet, contradictorily, by assuming he is identifying "light" with a single entity can Hanson make Newton's theory appear to be a straightforward dualist theory in any realist sense. Second, we know from Hanson's other writings that he is at least sympathetic to the view that terms common to distinct theories necessarily differ in meaning (see, especially, Hanson 1958). Hence we might want to claim on his behalf that his dualist interpretation is harmless since, according to him, 'wave' and 'particle' mean something at least slightly different in quantum mechanics as compared to classical physics. But this defense is misguided. So long as Hanson persists in identifying single *entities* as both waves and particles, as he does three times in the cited passage, he is assigning them incompatible properties, whether the terms are used either as they are in classical physics or in quantum mechanics.

The point I am making here is a simple but important one: even if we were to admit, strictly for the sake of argument, that the terms 'wave' and 'particle' have different meanings in quantum mechanics and in classical physics, merely because the two theories are different, the fact remains that the terms cannot be simultaneously attributed to a single entity. This is obvious in classical physics; as for quantum mechanics, we have seen that Bohr and Heisenberg, far from trying to break down the incompatibility, use it as the starting point of their analyses. Indeed, the whole point of speaking of "complementary but exclusive" descriptions is to emphasize the exclusiveness of the extended (ν, λ) wave description on the one hand and the localized (E, p) particle description on the other. The arguments which Bohr and Heisenberg use in the informal derivation of the uncertainty relations can be thought of as a proof of this exclusiveness.

It is important to realize that "dualism" in any realist sense is still unattractive if we take it to be making, not the absurd claim that a given item can be simultaneously wave and particle, but rather the claim that the item is appropriately described as a wave at one stage of a process and as a particle at another stage. For instance we might wish to say that in crystal diffraction experiments electrons are best represented as waves—as the actual oscillations of some physical item—while traveling from the source through the crystal lattice, but are best represented as particles when striking the detector.

Of course, in a very loose sense, this claim is true; but that is just the problem. If we take the waves to correspond to actual physical oscillations, then we are led to suppose an instantaneous contraction of the wave upon striking the detector, for no apparent reason. Similarly, if we assume the electrons leave the source because of localized energy and momentum transfer, then we are supposing that they leave as particles and suddenly, again for no apparent reason, expand into waves. Taken literally, such expansions and contractions seem to be magical events, allowing for no causal explanation. (Notice that what would apparently be involved here is a breakdown in *causality*, not merely a breakdown in *determinism*; the latter breakdown might well be taken in stride. See chap. 3 and, especially, chap. 4.) But if we are willing to countenance breakdowns in causality, then dualism becomes superfluous. We could get by with an explanation in terms of particles alone, if we assume that the crystal lattice scattering effects just happen to be such as to lead to an interference pattern. (Actually there is a plausible, unmagical explanation of the interference phenomenon in terms of particles alone that will be discussed below.)

Thus any interpretation which takes the quantum mechanical waves with ontological earnestness (takes them to be actual physical oscillations of some sort extended throughout space), but also takes quantum mechanical systems to be particles (local and stable concentrations of mass, charge, etc.), is either flatly self-contradictory or, at best, superfluous. The interpretation of Bohr and Heisenberg is not a dualist interpretation of either of these sorts. Rather, Bohr and Heisenberg take the fact that wave and particle properties are related in a surprising way, via equation (1), to indicate that neither kind of property should literally be attributed to physical systems at any time.

I have dealt at some length with three misconstruals of the Copenhagen interpretation because, unless the interpretation is attacked by refuting the actual arguments on which it is based, two very misleading impressions will be difficult to avoid. First, we will suppose that in refuting "subjectivism" or "positivism" or "dualism," in some realist sense, we will also be refuting the Copenhagen interpretation. Second, we will suppose that, having "refuted" the Copenhagen interpretation, a satisfactory alternative interpretation will be forthcoming if only we eschew positivism, dualism, etc. Neither of these suppositions could be further from the truth.

Genuine Objections to the Interpretation

Given that the above three objections to the Copenhagen interpretation are misdirected, any genuine criticism must be along different lines. The basic objection to the Copenhagen interpretation involves claims about the meanings of certain key terms common to classical and quantum mechanics. Accordingly, some preliminary remarks are in order on the notion of meaning, on criteria for meaning-change or meaning-invariance, and on the *classical* meaning of 'particle.'

The semantic aspect of the meaning of a term is closely related to those criteria which govern the "correct" usage (i.e., the standard usage) of the term by some group. When we ask what 't' means in this semantic sense, we are asking a question whose answer involves citing those properties an item or items have in virtue of which the term 't' is correctly and unequivocally applied to these items. In general, and this applies also in scientific discourse, there are several possible relations between terms and the criteria for correct application of these terms. However, one specific relationship is of particular importance for our discussion: the usage of some terms is such that there are logically necessary and sufficient conditions for correct usage. If an item lacking property P could not correctly be labeled 't' no matter what other properties the item possessed, then possession of P is a logically necessary condition for correctly labeling an item 't.' Conversely, if an item possessing property Q would automatically be labeled 't,' no matter what properties it lacked, then possession of Q is a logically sufficient condition for correct application of 't' to such an item.

In our criticism of the Copenhagen interpretation we will be exercised by the question of whether 'particle' can be construed to have the same meaning in classical and quantum mechanics. Given that we have specified the meaning of a term in an earlier theory, and wish to decide if the usage of this term in a newer theory constitutes a change in meaning, we have various clues or indicators, various criteria of meaning-change or meaning-invariance, to guide us. One clue is behavioral: if a term common to both theories is introduced in the newer theory without explicit definition or explanation of its usage, then, all other things being equal, we would expect some confusion or incomprehension on the part of readers or listeners if the meaning is different but not otherwise. Another clue involves the relative ease or difficulty of resolving apparent disagreements, or differences in the claims made using the common terms. If one theory appears to be asserting and the other denying that t's—items corresponding to 't'—have property P, then, if 't' means the same in both theories, there should be some way, in principle, to decide which claim is correct; there should be some theoretical or experimental evidence which is at least relevant, though it need not be decisive, in settling the apparent dispute. To the extent that scientists can agree on what is or would be relevant evidence for settling the disagreement, this is an indication that 't' is meaning-invariant. To the extent that there seems to be no way of settling the apparent disagreement, no relevant evidence which counts one way or the other, this is an indication that the meaning of 't' or 'P' or both is different in the two theories.

A third criterion for meaning-change or meaning-invariance involves the predictability of the application of a term in newer theoretical contexts on the basis of the criteria governing usage of the term in the old theory. If on the basis of the old use-governing criteria, we could naturally and plausibly predict what sorts of items—objects, properties, relations, etc.—the term would be applied to when using the apparatus of the newer theory, then this is some indication that the meaning of the term is the same in both theories; that understanding of the term in the new theory comes before, or at any rate is potentially independent of, understanding the new theory as a whole. Conversely, if the new applications of the term come as a complete surprise, if they are completely unpredictable on the basis of the old criteria, this is a strong indication that the meaning of the term has changed.

It is clear enough that in most modern textbooks of classical mechanics possession of simultaneous position and momentum is regarded as a logically necessary and sufficient condition for correct application of 'particle.' Although this condition is seldom stated in so many words, it is a trivial deductive inference from what is explicitly stated. One standard theoretical physics text tells us that bodies whose extension in space may be neglected may be represented as particles or mass-points (Joos 1934, p. 75). A second text tells us that particles are described when their position in space plus their mass and any other internal attributes—charge, spin, etc.—that are relevant to the particular problem at hand are given (Corben and Stehle 1960, p. 2–3). A third text defines particles as massive bodies of negligible dimensions (Becker 1954, p. 1). Of course, we are reminded by a fourth text (Landau, Akhiezer, and Lifshitz 1967, p. 3), what counts as a particle depends on the particular situation being analyzed; the earth is a particle, i.e., its dimensions are negligible, when we are interested in calculating its revolution about the sun but not when we are interested in calculating its rotation about its own axis.

The basic idea behind all these explications of the notion of a particle is that bodies whose mass (and other relevant internal attributes, if any) can be considered located at a single point at a given time, at a "center of mass," are bodies which are thereby represented or at least representable as particles. None of the cited texts mentions 'momentum' explicitly, but it is hardly necessary that they should do so. For it is obvious that the center of a mass of a body is either moving or not moving—with respect to an arbitrary coordinate system. Hence any massive body whose mass is locatable at a point, any mass-point or particle, will have velocity, zero or non-zero. Further, this velocity or position-derivative must be a single valued function of time, for if a body is moving with two (or more) different velocities at the same time then it is describable as two (or more) different particles.

So we see that possession of simultaneous position and velocity is a logically necessary and sufficient condition for correct application of 'particle,' according to modern classical mechanics texts. We could use this condition as the basis for our discussions, but to avoid confusion it is better to connect the condition with the term 'momentum,' which figures in the discussions of most other authors. The point to remember is that having the properties corresponding to 'mass' and 'velocity' is logically sufficient, though *not* logically necessary, for possessing (classical) momentum. The condition is not logically necessary because physical systems such as electro-magnetic wave-fields are said to have momentum even though they possess no mass and are not localizable. Examination of the usage of 'momentum' in classical physics reveals that items are said to possess momentum provided either they possess mass and velocity or are capable of changing the quantity of momentum of items which possess momentum. I emphasize again that possession of mass and velocity is *logically* sufficient for possession of momentum. If a physicist claimed that a body possessed both mass and velocity yet denied that the body possessed momentum, he would be regarded by his colleagues as contradicting himself.

Possession of simultaneous position and momentum is a logically necessary and sufficient condition for application of 'particle,' according to modern classical mechanics texts. It is this sense of 'particle' that we will claim can be carried over to quantum mechanics to provide a satisfactory interpretation of that theory. One aspect of this sense of 'particle' is particularly important: particles always travel in continuous paths—possession of unique velocity at all times is equivalent to possession of a continuous position vs. time curve.

Though it would be interesting to investigate the relation of this modern notion of 'particle' to the usage of 'particle' by earlier theoreticians, we shall not, for this is unessential for our purposes. It seems obvious that the pioneers of classical mechanics—Kepler, Huygens, Galileo, Newton, and others—possessed a concept of 'particle' closely related to if not identical with the standard modern notion, but we shall not rest any weight on this slightly controversial historical claim. The essential point is that we can and do understand any and all of the stages of classical mechanical development by approaching them through this modern notion of 'particle.' But can we also understand quantum mechanics in this way? Let us begin the long process of giving an affirmative answer to this question by turning to our criticisms of the Copenhagen interpretation.

The basic objection to the Copenhagen interpretation can be cast in the form of a dilemma. Bohr and Heisenberg have told us two things: physical systems cannot accurately be described as either waves or particles, i.e., cannot exactly satisfy any classical description; yet observed phenomena can and must be described classically. Another way of expressing this dilemma, of bringing out the tension between the two claims, is to use the formulation of it found in one standard textbook. Landau and Lifshitz (1958) tell us that: quantum mechanics requires us to give up the notion of the path of a particle; and that quantum mechanics is unique among physical theories in that "it contains classical mechanics as a limiting case, yet at the same time it requires this limiting case for its own formulation" (pp. 2–3).

We should not interpret their first claim to mean that although particles move in continuous paths, with simultaneous position and momentum values at all times, quantum mechanics does not and cannot enable us to calculate these individual paths but only to calculate other aspects of particle processes. This weak claim leads to no dilemma at all. It is obvious that Bohr and Heisenberg are making the stronger claim: that it is meaningless to attribute classical paths to physical systems. Landau and Lifshitz also intend this stronger interpretation, since they say that the results of electron diffraction experiments "can in no way be reconciled with the idea that electrons move in paths" and that "in quantum mechanics there is no such concept as the path of a particle" (pp. 2–3).

It might be supposed that the cited dilemma is a very superficial one and that a defender of the Copenhagen interpretation could easily extricate himself from it. After all, such a defender might argue, all we really mean when we say that quantum mechanics cannot be formulated independently of classical mechanics is that since quantum mechanical objects have no (exact) paths it would be impossible to formulate any mechanics for systems composed entirely of such objects—they would have no other (exact) dynamical characteristics either, if they have no (exact) paths; that what is required for a formulation of quantum mechanics is only the assumption that *some* physical objects obey classical mechanics *to a sufficiently high degree of accuracy*, although not exactly; and that hence the properties of quantum mechanical objects—e.g., electrons—can be inferred from, and only from, the changes they produce in classically described objects when they interact with these latter objects. This reply is entirely correct, but as a reply to the original dilemma it rather misses the point. The point is that, if quantum mechanics cannot be formulated without using classical concepts or terms, then those terms common to both classical and quantum mechanics must have the same meaning in both theories. Equivalently, we can say that, unless such common terms have theory-invariant meanings, any derivation purporting to show that classical theory is a special, limiting case of quantum theory would be invalid since it would rest on equivocations.

When an advocate of the Copenhagen interpretation claims that certain key terms—'particle,' 'position,' 'momentum,' etc.—cannot mean the same in quantum and classical theory, he cannot consistently go on to assert that classical theory is a special, limiting case of quantum theory. Bohr and Heisenberg do make both of these claims; hence the Copenhagen interpretation taken as a whole cannot be regarded as satisfactory. I emphasize that Bohr and Heisenberg do not intend to assert the relatively weak and uninteresting claim that the formalism of quantum theory includes that of classical theory. Rather, they acknowledge that classical concepts—i.e., terms with classical meanings—are absolutely essential for a description of experimental results, and that both theories make claims about a common set of items, a common set of referents for their common terms. Of course, we can hardly rest content with having shown that the two central claims of the Copenhagen interpretation are mutually inconsistent. A critic must also show that the spirit of the interpretation cannot be saved by relatively minor modifications. Finally, he must decide which of the two cited mutually inconsistent claims is true and which is false.

It is unsatisfactory to modify the interpretation so the claim that quantum mechanics includes classical theory as a special, limiting case is reconstrued to mean that the formalism of quantum theory includes that of classical theory. We cannot identify the theories in question with their respective formalisms since no formalism has a unique interpretation—the relation between theories and formalisms is not one-one. It is the interpretation of the formalisms in terms of some particular subject-matter which makes a theory a physical one as opposed to a mathematical or biological one. Moreover, when we are interested in coming up with an interpretation of a theory such as quantum mechanics, what we want to know is what the various terms of the theory mean. Typically, and historically this is certainly the case for quantum mechanics, we already know that the formalism of the theory is consistent and that the predictions of the theory agree with observation over a fairly wide range of phenomena. Hence we are not interested in the *syntactical* or grammatical usage of the terms—e.g., how to use them consistently in sentences or formulas—since we already know this usage and know that it does not uniquely distinguish the items corresponding to these terms. Rather, we are interested in the *semantic* aspect of usage—how to identify the items corresponding to a term, what counts as a correct instance of the terms, etc. An interpretation of quantum mechanics which confines itself to saying that the formalism of quantum theory includes that of classical theory, but that certain key terms common to both theories differ in meaning in the two theories, cannot account for the (approximate) agreement in the predictions of the two theories in the macroscopic range. To say that this agreement comes about *merely* because one formalism includes the other is nonsense, since we could associate either formalism with a set of claims concerning nonphysical matters—e.g., numbers and their relations.[3] Some minimal *semantic* interpretation of the two theories is necessary if an interpretation is to account for the cited predictive agreement.[4]

The most natural semantic approach is simply to say that the key terms common to both theories do mean the same and to show that apparent evidence to the contrary is misleading. To see which terms are involved, as well as to provide other background for ensuing discussions, it will be instructive to present an actual derivation, that shows classical mechanics to be indeed a special, limiting case of quantum mechanics.

For a system consisting of a single particle, with mass m, the time dependent Schrödinger equation can be written as

$$i\hbar \frac{\partial \psi}{\partial t} = \frac{-\hbar^2}{2m} \nabla^2 \psi + V\psi = H\psi, \qquad (3)$$

where $\psi = \psi(\mathbf{r}, t)$ is the quantum mechanical wave function; $V = V(\mathbf{r}, t)$ is the potential, classically described; $\hbar = h/2\pi$, and H is the Hamiltonian operator that contains information about the kinetic and potential energies of the system.

By inspection, the general solution of equation (3) can be written

$$\begin{aligned} \psi &= \text{const.} \times \exp[-(i/\hbar)\int H dt] \\ &= \text{const.} \times \exp(-iW/\hbar). \end{aligned} \qquad (4)$$

That is, $W = W(\mathbf{r}, t) = \int H \, dt$ is the quantum mechanical operator corresponding to the classical "action" function.

Substituting equation (4) into equation (3), differentiating and cancelling common factors yields

$$\frac{\partial W}{\partial t} + \frac{1}{2m}(\nabla W)^2 + V - \frac{ih}{2m}\nabla^2 W = 0. \qquad (5)$$

Now, if the kinetic and potential energies of the system and the time rate of change of the action of the system are very large compared to $\hbar$, even when this constant is multiplied by second (spatial) derivatives of the action function,[5] then the last term is negligible compared to the first three. Hence we have

$$\frac{\partial W}{\partial t} + \frac{1}{2m}(\nabla W)^2 + V = 0, \qquad (6)$$

which is the classical Hamilton-Jacobi equation for the "action" of a single "particle."

We can cast the above result in an even more familiar and useful form by making the classical identification of momentum with action gradient,

$$\mathbf{p} = \nabla W.$$

For then

Objections to the Copenhagen Interpretation

$$\frac{\partial W}{\partial t} + \frac{\mathbf{p}^2}{2m} + V = 0$$

or

$$\frac{\partial W}{\partial t} + H\,(\mathbf{p}, \mathbf{r}) = 0, \qquad (7)$$

where the classical Hamiltonian (H) is a function of $\mathbf{p}$, explicitly, and $\mathbf{r}$, implicitly, via the potential function (V).

Having sketched one version of the standard derivation of classical mechanics as a special, limiting case of quantum mechanics, we now want to know whether certain key terms common to both equations (3) and (7)—'position,' 'momentum,' 'mass,' 'potential,' 'time'—can be said to have the same meaning in both equations, or whether their meaning must be regarded as different in equations (3) and (7). Further, we want to know whether the appellation 'particle,' referring to the systems described by equations (3) and (7), must be said to differ in meaning, or whether the meaning of 'particle' can be considered invariant.

We must not allow our opponent to beg the question at this juncture; to say that, of course, the terms in question differ in meaning between equations (3) and (7) since every equation of every theory represents a logically necessary condition for correct application of its occurring terms, and hence the terms of equation (7) fail to meet the logically necessary condition of equation (3), since (7) implicitly denies what (3) asserts and vice versa. The *general* claim presupposed here by our opponent is refuted in chapter 3. Hence such an opponent cannot argue from the general claim to its application to the pair of theories classical/quantum mechanics. If the terms common to equations (3) and (7) are to be shown to differ in meaning, this showing must be on the basis of some criteria independent of the general claim. No such criteria can be found, since we can show that as actually used by physicists the common terms should be construed as meaning-invariant.

Let us consider each of the terms in question in turn. The derivation begins with the explicit assumption that the potential of the system is classically describable. Whether this assumption is ultimately correct is beside the point. What is relevant is that the derivation explicitly states that the description of the potential does not vary between equations (3) and (7); hence, according to the

35

derivation, 'potential' is to be regarded as meaning-invariant. Next, the property corresponding to the term 'mass' is tacitly assumed to (partially) characterize the type of system under discussion and to have nothing to do with differences in the dynamic behavior of the system, differences between the assertions of equations (3) and (7). In some texts this assumption is explicit, not merely tacit (Landau and Lifshitz 1958).

The claims here, so far unsupported, about the meaning invariance of 'mass' and 'potential' can be supported by explicitly invoking the criteria for meaning-change and meaning-invariance that we cited above. The terms 'mass' and 'potential' arc introduced in quantum mechanics without explicit new definition or qualifying explanation. Yet no one seems to be confused or dumbfounded by this practice; readers and listeners seem to understand well enough what is meant. They exhibit none of the usual behavioral signs of misunderstanding and incomprehension. Conversely, writers and speakers are never castigated for not defining these two terms. All parties seem to presume that the classical meanings carry over, and this presumption leads to no difficulties.

Specifically, there is no difficulty in showing the superiority of quantum mechanical predictions, involving these terms, over the corresponding classical predictions. Scattering cross-section predictions provide an instructive illustration of this point. Quantum mechanics predicts a value four times as high as the classical value for the total scattering cross section of low-energy particles from a rigid sphere, a situation closely approximated by the scattering of slow neutrons by very heavy nuclei. The two predictions use in their derivations statements employing the common terms 'mass' and 'potential'; indeed, these terms can be incorporated into the conflicting conclusions. That is, quantum mechanics asserts that for neutrons, of known mass m, the total scattering cross section by a sphere of radius a, where the potential is zero outside the sphere and infinite on and within the spherical surface, is $4\pi a^2$. Classical mechanics asserts the value πa^2 for a property described in exactly the same fashion. Theoretical and experimental physicists have no difficulty in citing evidence relevant to showing the superiority of the quantum mechanical prediction. (Of course, no great weight rests on the details of the particular example cited; it could be multiplied many times over and was chosen primarily because of its simplicity.) So, according to our two criteria for meaning-invariance, 'mass' and 'potential' would appear to have the same meaning in classical and quantum mechanics.

Having discussed 'mass' and 'potential,' we come to the controversial terms: the kinematic terms 'position' and 'time,' the dynamic term 'momentum,' and the descriptive label 'particle.' Strictly speaking, none of these terms are given explicit new definitions in quantum mechanics, but position and momentum are represented by noncommuting operators rather than by real-valued classical functions. That is, corresponding components of position and momentum have reciprocally related uncertainties or dispersions. (In nonrelativistic quantum mechanics time is treated exactly as in classical mechanics—as an independent functional variable; in relativistic quantum mechanics time is represented as an operator which does not commute with the energy operator.)

We should be careful not to make too much of this operator representation. Even in classical theory, position, momentum, etc., could be represented by operators; such representation is not in itself what distinguishes quantum from classical theory. Rather, it is the fact that conjugate variables do not commute that is distinguishing; equivalently, we could say that it is the uncertainty relations that are distinguishing. But this fact considered in isolation is most uninformative. Should we conclude from the cited fact that items labeled 'particles' do not have simultaneous position and momentum and hence that 'particle', 'position,' and 'momentum' cannot have the same meaning in equations (3) and (7)? We will show that this conclusion is too hasty, but before doing so let us reassure ourselves that the quantum mechanical properties of position and momentum are closely connected to familiar classical properties.

We have already noted that momentum is represented by an operator—proportional to the gradient vector operator—in quantum theory. We cannot compare operators with classical functions directly, but what we can and should compare with classical quantities are the expectation values, or most probable values, of the quantities represented by operators. It is easily shown that the expectation value of the quantum mechanical particle momentum is equal to the particle mass times the expectation value of the particle velocity. So to this extent at least the semantic aspects of 'momentum' are the same in both classical and quantum mechanics; in both theories the term refers, so far as particles are concerned, to the property corresponding to $m\mathbf{v}$. The difference is that classical mechanics presumes the dispersion in the expectation value can be made to approach zero; quantum mechanics denies that this can be done. The connection between classical and quantum mechanical *position* is even more direct and intimate; in both theories position is represented by the usual position vector. With this background in mind we can now see that the uncertainty relations—or, equivalently, the minimal reciprocal dispersions in the expectation values of position and momentum operators in quantum mechanics—provide no reason to suppose 'particle,' 'position,' or 'momentum' have different meanings in classical and quantum theory. / /

To see this it is convenient to discuss the term 'particle.' Possession of simultaneous position and momentum by an item is logically necessary for labeling that item a classical particle, according to the standard modern usage. Those philosophers who claim that "theoretical" terms, if not all terms, are "theory-dependent" will readily grant us this point. But it is important to remember that this condition is also logically sufficient. The central idea of classical mechanics is that bodies representable as particles move in continuous paths, with the correct geometric descriptions of these paths determined by laws of motion.

Clearly, if at all times a body possesses a position in space and, simultaneously, a unique instantaneous momentum, the body will move in continuous paths since, mathematically, the existence of such momenta—or, proportionally, velocities or position-derivatives—is equivalent to the existence of a continuous position vs. time curve.[6] To call an item a particle is to give a very minimal kinematic characterization of its properties. In calling the item a particle we are committed to saying that certain important properties of the item—the kinematic properties, position and velocity, and the dynamic property, mass—can be considered to be located at a geometric point. But we are not committed to saying that any *particular* set of laws of motion determines the particular continuous paths our particle or particles travel in. We are not even committed to saying that there are any laws which determine these paths. Different laws of motion would determine different paths, but all the paths would be continuous ones, this continuity being a mathematical consequence of the possession of simultaneous position-momentum values and hence a consequence of the minimal, formal description implicit in the term 'particle.' /

As to the second point, notice that the claim is there might be no laws which determine the paths in question—"determine" in the sense of uniquely predicting these individual paths. Of course if we are to have any reason to believe that such paths exist or, equivalently, that the items in question are particles, we must have some way of determining these paths. The fact is we can "determine" them, but only retroactively. We cannot predict them in advance nor can we use the retroactively inferred paths as the basis of unique predictions about the future.

To infer the simultaneous possession of position and momentum values, we merely perform the following thought experiment. Imagine a particle to pass through a very small aperture; the smaller the aperture, the smaller the uncertainty in the position of the particle when passing through the aperture. After the particle passes through the aperture, it will strike a detecting "screen" placed on the opposite side of the aperture from the source of the particle. We can then infer the momentum the particle must have had when passing through the aperture in order to arrive at the point at which it struck the detecting screen. There is no theoretical limit to the accuracy with which such simultaneous position-momentum values can be measured. But such a measurement is obviously retroactive, after the fact, and cannot be used as a basis for predictions. We cannot predict what the particle will do, if anything, after striking the screen since the position measurement implicit in its striking the screen will disturb the particle momentum.

Furthermore, we cannot infer anything about the paths of seemingly indistinguishable particles from the original one when these other particles are passed through the aperture. If the aperture is small enough, the probability of a particle colliding with the edge or rim of the aperture when passing through it is very high. In general, the position measurement will disturb the momentum of the particle. Hence we cannot be said to have measured the simultaneous position and momentum that the particle had before passing through the aperture. Accordingly, the paths of similar particles cannot be individually predicted. Once the particle hits the screen, however, there is a clear sense in which we have measured the simultaneous position and momentum at the instant after passing the aperture.

The above "experiment" raises a terminological question. We might claim that the concept of measurement involves some notion of at least approximate reproducibility, and so the above-described procedure should not count as a genuine measurement. There is something to this claim, vague as it is, but the important point to notice is that we need not label our procedure a 'measurement,' in this strong sense, in order to establish the point of interest—we have shown that there is reason to believe that each individual particle has simultaneous position and momentum. Accordingly, there is reason to believe that those individual particles trace out continuous paths in space. It is popularly supposed that this conclusion is ruled out by the uncertainty relation between position and momentum, or, equivalently, by the wave functional description implicit in equation (3). To see why this supposition is mistaken, we must examine the application of equation (3) to specific problems, the most instructive of which are scattering problems.

In solving a scattering problem—e.g., in predicting the energy and angular dependence of a beam of particles scattered from a target—the quantum theorist solves equation (3) to obtain the wave function after the scattering events, and interprets $|\psi|^2$, the square of the amplitude of this scattered wave function, as giving the relative probability of scattering through a given angle. That is, equation (3) correctly predicts the relative number of particles from the original beam which are scattered through various angles. The correctness of such predictions is inferred from the number of *localized interactions* of the particles of the scattered beam that are registered on suitable detecting devices. //

This brings out two points. First, both the particles of the beam and the classically described macroscopic detecting devices are regarded as having simultaneous position and momentum, within the limits of the uncertainty principle. Second, we now realize that equation (3), although originally described as the equation for a single particle, is actually interpreted, by theoretical and experimental physicists, as describing the statistical behavior of a beam of particles. But why can we not interpret equation (3) as an equation for a single particle? Well, in a way we can—but only with a little care. Suppose we try to make such an interpretation by considering a scattering experiment involving a incident beam consisting of a single particle. Of course as a practical matter such beams are unproducible, so this must be regarded as a thought experiment, but this fact does not affect the point under discussion.

The point is that the scattered wave function for a single-particle incident beam will in general have nonzero values for all angular directions; but the scattered "beam" will be observed to strike a detector oriented at one particular angle. What relation is there, then, between the predicted angular distribution and the observed direction of scatter? None at all—the scattered particle will strike a detector at some angular orientation but we cannot be said to have predicted this event from equation (3), since what (3) predicts is an angular distribution. So if we are to regard equation (3) as correctly predicting the behavior of a single particle, the only thing this can sensibly be taken to mean is that, if the original particle were to be somehow *repeatedly* brought back into the experimental arrangement as a one-particle incident beam, the observed results of successive scattering events involving this same particle would agree with the angular distribution predicted by equation (3). These facts bring out the important point that equation (3) and the uncertainty relation implicit in it do not describe the individual paths of individual particles but rather describe either: the statistical distribution of simultaneous paths of a beam of similar particles; or the statistical distribution of successive paths of the same single particle in repetitions of the same experimental arrangement.

With the above points in mind we can now see that the Copenhagen interpretation is too strong, or at the very least too hasty, when it claims that we cannot interpret subatomic systems as particles or when it claims that we must give up the concept of the path of a particle. On the contrary, as we have seen, physicists do interpret such systems as particles, and an examination of their actual procedure has indicated that, if this pocedure is to make any sense, we must regard them as implicitly assuming that 'position,' 'momentum,' etc. have the same meaning in both classical and quantum mechanics.

Let us briefly reiterate the basis for these conclusions. We have seen that physicists do attribute to systems called 'particles' simultaneous position and momentum values, within the limits of the uncertainty principle. When they measure scattering cross sections by counting localized interactions of particles with detectors, they are attributing to the particles a definite position at the instant of impact with (some small region of) the detector; further, they attribute momentum and energy to the particle at this same instant

—it is the momentum and energy which produce the noticeable changes which eventually show up as detector counts.

Finally, we have seen that the uncertainty relation between position and momentum components does not preclude simultaneous possession of position and momentum by these particles. The uncertainty relation always refers to ensembles of particles— many particles at the same time or the same particle (i.e., indistinguishable particles) at many different times; the relation could not sensibly be taken to refer to the individual paths of individual particles. Putting these results together, we see that 'particle' can be construed as meaning the same—as referring to items which necessarily possess simultaneous position and momentum—in classical and quantum mechanics. As for 'position' and 'momentum' themselves, their meaning-invariance can be inferred from our previously cited criteria. In the absence of explicit new definitions, the lack of confusion or incomprehension and the apparent ease of deciding the relative superiority of quantum mechanical over classical claims formulated in terms of 'position' and 'momentum' count in favor of meaning-invariance. The same argument given for 'mass' and 'potential,' an argument involving scattering cross-section predictions, could be repeated, change for change, for 'position' and 'momentum.' Finally, the fact that 'position' and 'momentum' pick out the same properties, the position vector and the quantity $m\mathbf{v}$, respectively, counts in favor of meaning-invariance. It is plausible and natural to suppose that one familiar with the classical criteria governing usage of 'position' and 'momentum' would be able to predict the quantum mechanical applications of these terms, given that they are to be represented by noncommuting operators.

Furthermore, to say that we must give up the concept of the path of a particle is a very misleading way of saying we must—on present evidence—give up the belief that individual paths of individual particles are predictable. This latter way of putting the matter indicates quite clearly what it is which must be given up— determinism, and not the concept of the path of a particle.

Some might object to this trade-off of determinism for preservation of the concept of the path of a particle. After all, an opponent might argue, in the present context to say that we need "only" give up determinism means that we need to give up belief in the exact correctness of laws predicting future simultaneous position-momentum states from earlier position-momentum states; to give up this belief because such states cannot be exactly specified at any time. Hence, the argument goes, in giving up this aspect of determinism we are giving up that classical description whose satisfiability is required for correct application of the term 'particle.' But, once again, anyone who argues in such fashion begs the question. We have seen that physicists employ the term 'particle' and the terms for various particle properties—'position,' 'momentum,' etc. —in such a way as to reveal that these terms have a common meaning in classical and quantum mechanics. An exact classical description does not specify necessary conditions for correct application of the term 'particle.'

The point of view that I have been urging with regard to the term 'particle' amounts to the following. The central idea of classical mechanics can be separated into two ideas—first, that bodies are representable, in certain contexts, as particles, i.e., as having simultaneous position and momentum and hence as traveling in continuous paths; and second, that these paths are determined by a particular set of laws of motion. Once we notice this separability we see that the essential core of the concept of a particle can survive in going from a theory embodying one set of laws of motion to a theory embodying another set; the bodies whose motions are predicted by the new theory can still unequivocally be called particles, in certain contexts, since they have simultaneous position and momentum and hence travel in continuous paths. Furthermore, even if there is no set of laws of motion which determine these paths, the bodies can still unequivocally be called particles provided there is reason to believe they have simultaneous position and momentum.

The difficulty in showing decisively that physicists would be willing to apply the term 'particles' to objects described by nonstandard but still deterministic laws of motion lies in the seemingly counter-factual character of this test case. It would appear that, historically, there has been only one (fully developed) deterministic mechanics. Hence before quantum mechanics, there had never been any clear need to separate the central idea of classical mechanics into two parts. But once this separability is noticed, we can understand why physicists have in fact continued to label as 'particles' such objects as electrons, protons, neutrons, etc., even though they have abandoned the belief that there are any laws of motion which exactly determine the trajectories of these quantum-mechanical objects. That is, they have implicitly accepted the separation I have noted and have indicated, by their linguistic practice, that consciously or otherwise they feel only the first part of the cited central idea of classical mechanics is the essential core of the notion of a particle.

Furthermore, the monolithic appearance of classical mechanics is somewhat misleading. Actually 'classical mechanics' covers three quite different historical periods. In the first, pre-Newtonian period, scientists such as Kepler and Galileo discovered special laws describing the behavior of (objects representable as) particles. In the second period Newton discovered supposedly general laws governing all (particle) motions. But these Newtonian laws were eventually found inadequate, even apart from quantum mechanical considerations. That is, Newton's laws do not transform properly, according to relativistic requirements, even when we are careful to avoid the nonrelativistic assumption that mass is independent of velocity. For when electromagnetic interactions are involved, we must replace the usual momentum $\mathbf{p}$ by

$$\mathbf{p}_{em} = \mathbf{p} + \frac{e}{c}\mathbf{A}$$

for particles of charge e and for electromagnetic vector potential $\mathbf{A}$. It is this revised $\mathbf{p}$ which must be used in the usual force law, i.e., in Newton's second law.

Accordingly, we can now see how we can use the term 'particle' as we do in fact use it, in a theory-invariant way, in interpreting all three stages of development of classical mechanics. If we presume that the criteria governing usage of 'particle' are the same no matter which of the stages we are discussing, we shall have no trouble understanding any of the stages in isolation or understanding their mutual relations. In particular, if we assume possession of simultaneous position and momentum to be logically necessary and sufficient for correct application of 'particle,' then surely no one will be confused or dumbfounded by this usage. We shall have no difficulty, for example, agreeing on the relevant evidence for showing the superiority of Newtonian over Galilean or Keplerian predictions. Indeed, this is no mere hypothetical exercise of imagination; it closely resembles actual discussions sometimes given in textbooks. Hence, according to our criteria for meaning-change, it is most implausible to deny the possibility of the meaning-invariance of 'particle' throughout these classical developments.

Once we appreciate the classical situation with respect to the term 'particle,' then a certain use of the term in quantum mechanical contexts is to be expected, a use where the criteria governing application are the same as those specified for the classical usage. That is, once we see that the meaning of 'particle' is not necessarily tied to that set of dynamical laws regarded as correct at a given time, then we see that the term can be applied without equivocation in a situation where theoreticians deny that there are any (exact) dynamical laws. This usage is not only possible, it is plausible and natural. Indeed, no alternative usage seems to be ultimately satisfactory, which is just a way of saying that no alternative interpretation of quantum mechanics is satisfactory—a claim to be substantiated in later chapters. Before leaving the present chapter, let us reassure ourselves that our conclusions to date square with our conceptual intuitions.

The best way of accomplishing this task of reassurance and at the same time of casting further doubt on the Copenhagen interpretation is to apply a quantum mechanical description to a familiar macroscopic process. In explicating the Copenhagen interpretation, we saw that Bohr and Heisenberg clearly chose to say that the quantum mechanical differences between atomic and macroscopic phenomena are differences of degree, not of kind. It is worthwhile to notice why the alternative choice is not open to them or to us. There are two ways of doing this.

First, the derivation sketched above indicates that the quantum mechanical description is the more general, that it is intended to apply everywhere—over the entire range of values of dimensions, momenta, etc. which physical systems can have. The classical description is exactly correct only in the limit of infinite momenta, a limit which actual physical systems never, quite, reach. Further, the more general quantum mechanical description has been found to be in better agreement with experimental observations than the more limited classical description over a wide range of, say, momentum values. But for very large bodies and/or bodies with high momenta both the classical description and the quantum mechanical description are correct to well within experimental error. What then should we conclude about the relative superiority of the two descriptions in the range where the differences between their predictions are undetectable? To assume that the relative superiority of the more general quantum mechanical description persists into the range of undetectable differences is merely to make an extrapolative or inductive inference. To suppose that the relative superiority is reversed in this range of undetectable differences is to make a counter-inductive inference. Once we realize which inference is inductive and which counter-inductive we need no longer wonder which inference is better supported by the evidence, since the answer to this question is obviously implicit in the meanings of the terms 'inductive' and 'counter-inductive.'

Second, we can arrive at the same result by consideration of the uncertainty relations—equation (2)—and the "dualist" relations—equation (1)—from which they can be derived. This second procedure is really equivalent to the first, since the relations of equation (1) are built into the quantum mechanical description which equation (3) represents. The point here is that we cannot suppose that equations (1) which, ultimately, prohibit simultaneous specification of position and momentum apply only to atomic or subatomic particles. The application of the equations could not be limited to such particles and also be limited to any particular momentum range. Subatomic particles, or at least charged ones, can in principle be accelerated to any desired momentum value. There is nothing in either classical or quantum theory to prevent this. Although the rest mass of such particles is small and their velocity is limited by that of light, any desired momentum value can, in principle, be achieved because of the increase of mass with velocity. So we must conclude that quantum mechanical effects apply to momenta in the range of values possessed by macroscopic objects as well as those typically possessed by atomic and subatomic ones.

As an example of a macroscopic phenomenon consider the trajectory of a baseball after being struck by a bat. In what sense is a particle description of baseball trajectories precluded by the knowledge that the uncertainty relations apply to baseballs? It is clear that we regard a material object such as a baseball as occupying some volume of space; as having mass, which for many purposes can be regarded as concentrated at a "center of mass"; as having a momentum value at various times; and as having a location, specified by that of its center of mass, at various times.

We now know that, contrary to our prequantum mechanical intuitions, we cannot, in principle, specify exact simultaneous values of the position—of the center of mass—and the momentum of the baseball, even though these theoretical uncertainties are far smaller than purely experimental errors. In view of this new knowledge should we now, in the spirit of the Copenhagen interpretation, regard baseballs as items without any intrinsic properties, as items which are no more particles than waves? Surely not. We still regard baseballs as particles in the following sense: they are entities which carry momentum and are capable of exchanging part or all of it; they have location at all times.

The claim that these gross particles have exact location at all times is not precluded by the uncertainty principle. We can always determine the location of the center of mass of a baseball provided a corresponding momentum uncertainty is accepted. Furthermore, it is obvious that this center of mass is located inside the surface of this extended, three-dimensional distribution of mass. Now consider the claim that a baseball, because of the now familiar arguments, *has* no simultaneous position and momentum at any time. The first thing to note about the phrase 'has no simultaneous position and momentum' is that it is a particularly tricky and ambiguous one. If the phrase means that the particles in question do not have these "conjugate" properties in the sense that the individual motions of individual particles cannot be traced in continuous paths, according to the equations of quantum theory, then of course the claim is true. But we have shown that the particles possess simultaneous position and momentum, provided we are willing to admit random, unpredictable changes in the individual motions of individual particles upon interaction with other systems, since such changes would account for the inability of (any) theory to trace individual paths in an exact fashion.

Once we realize this it becomes quite counter-intuitive to refuse to admit that baseballs are particles, given that they satisfy the criteria elaborated above. We cannot here be accused of equivocating on the term 'particle' since, as argued above, we have seen that the meaning of this term is invariant in classical and quantum mechanics. What the present argument amounts to is that it is sufficient for correct application of the term 'particle' that an item such as a baseball have an obviously noticeable momentum and a location characterized by that of its center of mass—a location which could not be other than inside the surface of the baseball. A relatively local and stable concentration of mass with such position and momentum properties can surely be correctly labeled a particle without doing violence to our classical intuitions. To suppose otherwise is to erroneously suppose the equations of classical mechanics—with its built-in determinism—specify necessary conditions for our conception of particle.

Once the above points regarding baseballs are accepted the implausibility of the Copenhagen interpretation becomes apparent. The interpretation is only plausible for unfamiliar, not directly observable, atomic entities. Once we realize that the difference in quantum mechanical effects between these atomic entities and macroscopic ones is a difference of degree and not of kind then the interpretation loses all persuasiveness.

It will be instructive for us to compare quantitatively the relative uncertainties associated with the paths of subatomic items on the one hand, and macroscopic items on the other. From equation (2) we know that given a position measurement accurate to within 10^{-8}cm—the same order of magnitude as an atomic diameter—the corresponding velocity uncertainty for a particle of mass m is given by

$$\Delta v \cong \frac{h}{m\Delta r} = \frac{6.62 \times 10^{-27}}{10^{-8}m} = \frac{6.62 \times 10^{-19}}{m}.$$

For a particle such as an electron, $m \cong 1 \times 10^{-27}g$ so that $\Delta v \cong 7 \times 10^{8}$cm/sec. In order to determine that an electron is within an atom at all, which is not to determine much on an atomic scale, we cannot infer from our measurement whether the electron is standing still or traveling 10,000 times as fast as a typical rifle bullet. For a particle such as a baseball, $m \cong 150$ g so that $\Delta v \cong 4 \times 10^{-21}$cm/sec. This latter uncertainty is far below the sensitivity of any modern measuring instruments, a fact which we should always keep in mind if we want to view the uncertainty principle in proper perspective. In fact, if we relax the precision of our position measurement on the baseball from the unattainable (from a practical point of view) value of 10^{-8}cm to a relatively sloppy value of, say, $\Delta v \cong 1 \times 10^{-2}$cm, the corresponding velocity uncertainty of $\Delta v \cong 4 \times 10^{-27}$cm/sec is even further below achievable precision of modern instruments.

One final way of rendering suspect the Copenhagen interpretation of Bohr and Heisenberg is to notice the untenability of the complete analogy between material particles and photons which they rely on. It is simply false, contrary to their occasional claims, that individual material particles can only be correlated with the intensity of the associated wave function, as is the case with photons. The path of individual particles can, at least approximately, be followed, e.g., in cloud chambers. The path of an individual photon cannot be followed at all—we can only infer the existence of a photon or photons at a momentum-energy transfer point by noting the amount of transfer and computing the number of discrete units transferred. How much can be made of this difference depends on considerations involving the quantum theory of fields, the discussion of which must be postponed until chapter 7.

It is revealing that Bohr himself, in a slightly later paper than the one under discussion, indicates some awareness of and sympathy for the point of view I have been urging above. Thus:

It is, after all, one of the most peculiar features of quantum mechanics that, in spite of the limitation of the classical mechanical and electromagnetic conceptions, it is possible to maintain the conservation laws of energy and momentum. In certain respects, these laws form a perfect counterpart to the assumption, basic for atomic theory, of the permanence of the material particles, which is strictly upheld in the quantum theory even though the conceptions of motion are renounced (1961, p. 113).

Furthermore, in a passage immediately preceding the quoted one, he mentions the important difference between photons and material particles cited above.

To have pointed out difficulties in the Copenhagen interpretation, as we have done, is far from having advanced a better interpretation, or any interpretation at all. As is obvious, I intend to defend an indeterministic particle interpretation of nonrelativistic quantum mechanics—the theory covering phenomena in which electromagnetic radiation is not involved and in which velocities are small compared to that of light. Since it is far from obvious that the problems of the quantum theory of material particles and that of electromagnetic fields can be discussed in isolation, a later chapter will deal with field considerations. Apart from this problem of fields, there are other problems facing any indeterministic particle interpretation.

The remainder of this essay consists of an explication of the precise sense in which quantum mechanics is an indeterministic theory; of why various deterministic reconstruals are misguided; and of analyses which reveal that phenomena popularly alleged to be recondite are amenable to particle explanations.

3

The Indeterministic Character of Quantum Mechanics

We could approach the philosophical issues of determinism by presuming the claim that physical processes are deterministic is necessarily true or necessarily false and advancing a priori arguments to support one or the other of these two conclusions. This procedure is quite suspect to anyone of empirical outlook, and the inconclusiveness of traditional philosophical debates on determinism reinforces our suspicion. Recently, a more tentative approach to the issues of determinism has been pioneered by various authors, notably Reichenbach (1946), Margenau (1950), and Nagel (1961). These writers suggest that we conceive determinism, or indeterminism, as a "structural" feature of particular physical theories and only by extension as a feature of those events and processes in the world which a theory is intended to describe, predict, and explain. Thus our opinions as to the deterministic character of the world may change as our knowledge changes—as older theories are replaced by newer, better established theories that may differ from the older ones with respect to this "deterministic" feature.

Although the cited authors share a common approach, their results are strikingly different. Reichenbach concludes that quantum mechanics, the presently best-established theory about all physical processes, is indeterministic. Margenau and Nagel argue that Reichenbach's conclusion is hasty since he presumes that a theory must be deterministic with respect to the specification of those state variables used in classical mechanics or not at all. Those two writers then go on to claim that quantum mechanics is deterministic because it satisfies a criterion for deterministic theories which is alleged to be a natural extension of that applicable to classical mechanics. In this chapter I show that their proposed criterion rules out no theories and hence completely trivializes the notion of determinism. I will cite specific respects in which the criterion simply is *not* an extension, natural or otherwise, of that applicable to classical theories.

A brief preliminary remark is in order on the terms 'determinism' and 'causality.' Both terms are used in a wide variety of ways but such variety need not be confusing provided we specify in each context the intended sense. We should avoid the casual identification of the two concepts. Throughout this chapter the issue in question is whether quantum mechanics is an indeterministic theory. Supposing the answer is yes, we cannot then conclude that the theory therefore involves or countenances breakdowns in causality. The events described by the theory are causal in the sense that later physical states arise via transitions from earlier ones and, in general, these transitions are due to disturbances capable of producing them. Even in those cases where isolated systems undergo transitions in the absence of external influences, the transition process is causal in the sense that the earlier state contained sufficient internal energy, momentum, etc. to account for the values of these quantities in the later states. That is, a theory is a causal one if the processes it describes satisfy momentum and energy conservation laws. But it is deterministic only if it is capable of establishing one-one rather then one-many correspondences between the earlier and later physical states involved in these processes.

Nagel proceeds (1961, p. 295) by setting up and attacking the following popular argument: the uncertainty relations assert that the position and momentum of a particle are not simultaneously ascertainable with unlimited accuracy; indeed these two parameters are not independent of each other but are so related that sharply defined spatial location is incompatible with sharply defined momentum; hence the equations of quantum mechanics cannot establish a unique correspondence between precise position and momentum values at one time and precise position and momentum values at other times. Nevertheless quantum theory is capable of calculating the *probability* with which a particle has a particular position when it has a particular momentum, and vice versa; hence quantum mechanics is not deterministic but rather irreducibly "statistical" in character.

Nagel attempts to subvert the above argument by denying that the term 'particle' and the terms 'position' and 'momentum' could possibly mean the same thing in quantum mechanics as they do in classical mechanics, claiming that the equations of quantum theory or indeed any theory "implicitly define," in a sense discussed below, the elements and processes postulated by any "model" for the theory (pp. 300–301). Accordingly, these elements must at least have the structural characteristics stipulated by these equations, including both the axioms and theorems of the theory. In particular, any model for the theory must satisfy the uncertainty relations. In classical mechanics the terms 'position' and 'momentum' satisfy different axioms (and hence different theorems) from those of quantum mechanics. These terms are used in such a way that a particle must always have a determinate position and simultaneously a determinate momentum. The uses "legislated" (Nagel's word) for these terms in quantum theory are, as indicated above, quite different. Hence, Nagel concludes, 'position' and 'momentum' cannot have the same meaning in the two theories.

The force of Nagel's argument depends on the claim that every axiom and every theorem of a given theory provides a logically necessary condition for correct usage of the terms occurring in these axioms and theorems.[1] This claim immediately leads to incredible consequences. To see this, let us examine the actual axioms of classical mechanics, which may be stated mathematically as

I   (i)   If $\mathbf{F} = 0$, $\mathbf{a} = 0$ or $\mathbf{v} =$ constant

    (ii) $\mathbf{F} = m\mathbf{a} = \dfrac{d\mathbf{p}}{dt}$

    (iii) $m_x\mathbf{a}_x = - m_y\mathbf{a}_y,$

where $\mathbf{F}$, $\mathbf{p}$, m, and $\mathbf{a}$ stand for 'force,' 'momentum,' 'mass,' and 'acceleration,' respectively, and the boldfaced quantities are vectorial —possessing direction as well as magnitude; the subscripts x and y designate two distinct bodies isolated from any third body. The theorem expressing the conservation of momentum is an immediate consequence of these axioms.[2]

Now suppose, counterfactually, that we had discovered that subatomic "particles" obey the following laws of motion:

II (i) If $\mathbf{F} = 0$, $\mathbf{v} = 0$ or $\mathbf{r} = $ constant
   (ii) $\mathbf{F} = m\mathbf{v} = \mathbf{p}$
   (iii) $m_x\mathbf{v}_x = -\,m_y\mathbf{v}_y,$

or these laws:

III (i) If $\mathbf{F} = 0$, $\mathbf{j} = 0$ or $\mathbf{a} = $ constant
   (ii) $\mathbf{F} = m\mathbf{j}$
   (iii) $m_x\mathbf{j}_x = -\,m_y\mathbf{j}_y,$

where $\mathbf{r}$ is the position vector and $\mathbf{j} = \dfrac{d\mathbf{a}}{dt}$ is the "jerk" vector or time derivative of the acceleration. In the case of equations II we would find that the quantity $m\mathbf{r}$ rather than $m\mathbf{v}$ would be conserved. Should we now conclude that the theories characterized by II and III are using the terms 'position' and 'momentum' with different meanings from their meanings in classical mechanics (I)?[3] Our first reaction to this question is that unless the kinematic terms 'position,' 'velocity,' 'acceleration,' and 'jerk' are used with the same meaning in all three theories we would have no way of finding out which of the three theories, if any, was correct. And, of course, in fact such kinematic terms are defined independently of, and at least to a good first approximation can be measured independently of, any particular dynamical theory, I–III or other.[4] If this were not the case, no observationally based evidence about positions, velocities, etc. could enable us to decide which dynamical theories, which laws of motion, I, II or III, were true and which false. What this exercise of imagination shows is that Nagel's explicit criterion that every theoretical axiom and theorem is a logically necessary condition for correct application of the occurring terms is much too strong. This must be the case at least for kinematic as contrasted with dynamical terms.

Since my claim is of pervasive importance in the philosophy of physics, it is worthwhile to elaborate on the basic objection to Nagel's position and to consider some possible counterarguments. What the objection amounts to is that unless at least some of the terms occurring in each of the two or more different theories had the same meaning these theories could neither agree or disagree in the claims they make about the items (objects, properties, relations) which the terms stand for. Thus unless 'position,' 'velocity,' and other kinematic terms have the same meaning in I–III then when theory II claims that it is the quantity $m\mathbf{r}$ rather than momentum which is conserved, or when theory III claims that $m\mathbf{j}$ is the conserved quantity, then theories II and III are not really denying anything which theory I asserts, since an equivocation of meaning is involved.

It is certainly counterintuitive to maintain that theories never contradict each other. Part of our notion of theories is that they do make conflicting claims. Another incredible consequence is that equations from two or more distinct but noncompeting theories could never be used within the same derivation if these equations contained any common terms. Again, the conclusion of such a derivation would rest on an invalid argument, one involving equivocations. But such derivations are common in physics and are not considered invalid, which indicates that the meaning of the common terms is considered invariant in the different theories by the users of these terms. An obvious example is the derivation of the equation of motion of a charged body moving in a variable electric field. This derivation uses equations from both mechanics and electrodynamics, equations containing common terms, notably 'position,' 'force,' and, of course, 'time.' These considerations show that Nagel's general claim about the necessary meaning-variance of theoretical terms involves consequences which are quite counterintuitive; the claim, therefore, should be regarded with utmost suspicion. Subsequently, we will see that there is no way that the general claim could coherently be established.

Still more unpleasant consequences face anyone who holds—Nagel does not, though some contemporary philosophers of science do[5]—that *all* terms which scientists use have necessary conditions for their correct usage specified by all the axioms and theorems of a given theory in which they appear. That is, if even the terms used in reporting observational findings were dependent for their meaning on—i.e., had necessary conditions for their correct usage specified by—all the theoretical axioms and theorems in which they occur, then no observational evidence could ever conflict with the theory. Theories would be analytic, necessarily true in virtue of the meanings of the terms involved. Nor could any evidence be provided, without begging the question, that some actually existing item satisfied the axioms of the theory, since any report of the observation of the existence of such an item would use the terms of the theory and these could not be used correctly except if satisfying the axioms. Theories could neither be falsified nor confirmed by independent evidence. Again, this is an absurd consequence, inconsistent with our notion of what theories are. Finally, we could not even learn to understand a given theory by having it explained to us in any terms whatever—the terms of the theory itself would not do since these can only be understood after the theory is understood; but any other terms would be irrelevant since their connections with the theoretical terms would be unspecified.[6]

Nagel does not hold that *all* terms are completely theory dependent and hence escapes the objections cited in the immediately preceding paragraph. Rather, Nagel makes a radical distinction between *theoretical* terms, which are completely theory dependent in the sense under discussion; and *observational* terms, which are completely independent of any theory and whose meaning is exactly the same in all theories. Whatever meaning theoretical terms have is alleged by Nagel to be given by "coordinating definitions" which are statements—equations, usually, in the case of physics—containing both theoretical and observable terms and hence providing a "partial" or "indirect" interpretation of these theoretical terms. Nagel also distinguishes between the *theory proper*, statements containing only theoretical terms; *coordinating definitions*, statements containing both theoretical and observable terms; and a *model* or interpretation of the theory proper in terms of familiar, macroscopic items (objects, processes, relationships, etc.).

These distinctions of Nagel hardly enable him to establish his general claim that each theoretical axiom specifies a logically necessary condition for correct application of its occurring terms.[7] For by any sensible standard, the kinematic concepts in question ('position,' 'velocity,' etc.) are either observable terms or explicitly definable in terms of observables. To avoid admitting this, Nagel would claim that these terms cannot mean the same thing when applied to unobservable, subatomic particles as when applied to macroscopic objects. This claim is both extremely dubious and, more important, not directly relevant, since by hypothesis we are considering three different theories (I–III) all about subatomic particles. So now Nagel must claim that the meanings of the kinematic terms differ in theories I to III because they are connected to observables by different sets of coordinating definitions. But to argue in this way is to give up the original criterion.

This defense now amounts to saying that common theoretical terms occurring in distinct theories cannot have the same meaning because their connections with observational consequences are different. Before the coordinating definitions which determine these consequences are specified, the terms have no meaning at all, according to the presupposed Nagelian distinctions. Thus the "defense" under consideration is incoherent if it argues both that: (a) theoretical terms unconnected to observables have no meaning; and (b) theoretical terms common to distinct uninterpreted—in the same sense given in (a)—theories necessarily differ in meaning. It is incoherent to argue that two or more terms necessarily differ in meaning if they have no meaning. A defender of the original criterion must ignore terms in their uninterpreted status. It is equally absurd to argue that terms common to distinct theories necessarily differ in meaning because the theories are necessarily connected to observables via different coordinating definitions.

It is easy to generate counterexamples to the claim. Consider the case where the sentences of two distinct "theories proper," sentences containing only theoretical terms, differ, but the two sets of sentences expressing the "coordinating definitions" are the same. For example, we can imagine alternative kinetic theories, in one theory the individual molecules obeying classical mechanics, in the other not; but in both cases the coordinating sentences linking properties of molecules (momentum, kinetic energy, etc.) to macroscopic properties (pressure, temperature, etc.) of gasses are the same. Here it is obvious that alleged differences in the meaning of theoretical terms cannot be attributed to differences in coordinating definitions.

Thus Nagel's distinctions in no way enable him to escape the original criticisms made above of his criterion. So long as no coordinating definitions are presumed specified, he cannot coherently assert the necessary difference in meaning between theoretical terms common to distinct theories. But the mere existence of these coordinating definitions does not enable him to handle the above counterexample and others like it. Nor will it help to say that the coordinating definitions are "analytic," each specifying a necessary condition for correct application of its occurring terms. Either the coordinating definitions are stipulated or discovered. If the former, they can be stipulated to be the same; if the latter, we cannot rule out in advance that they will be discovered to be the same. The conclusion that the two sets of coordinating definitions *must* differ cannot be reached.

Finally, consider the rejoinder that theoretical terms unconnected to observables are not devoid of meaning but rather are implicitly defined by the sentences of the theory—i.e., the "theory proper." Nagel's notion of an implicit definition is a curious one. What he means by an "implicit definition" is a purely *syntactical* definition, one which distinguishes the implicitly defined terms by means of the logical or formal structure of the theoretical sentences in which they occur.[8] The point we must remember here is that such formal structures never distinguish the meaning of their occurring terms from the meaning of other terms, terms occurring in theories with the same formal structure but obviously distinguishable on other grounds. We know, from a previously cited meta-mathematical theorem, that every consistent formal calculus has an interpretation in the realm of natural numbers.[9] The purely formal calculus of, say, classical mechanics no more distinguishes the meaning of its occurring terms from the meaning of corresponding terms in quantum mechanics than it distinguishes the latter theory from some set of arithmetic statements.

Thus, to say that a theory's formal structure uniquely distinguishes the meaning of its occurring terms is demonstrably false; to say that this structure distinguishes these meanings at all is a misleading way of saying that the theoretical claims concerning the items corresponding to the theoretical terms differ from the claims of some other theory dealing with the same subject matter.[10] To say that the "implicit definition" of the terms of classical mechanics confers *meaning* on the terms of the theory is odd indeed when we realize that such a "definition" does not enable us to distinguish the items corresponding to these terms from such things as numbers, arithmetic operations and relations, etc. Hence, although an implicit definition does tell us something about the grammatical and/or syntactical usage of a term (how to use it consistently in sentences), it tells us nothing about the more important semantical usage (about what, or even what sort of, non-linguistic item the terms corresponds to). But in interpreting physical theories it is precisely this semantical aspect of usage that we are interested in.

So far we have seen that Nagel's criterion cannot be generally correct. Terms common to two or more distinct theories can and frequently do have the same meaning. The moral of the foregoing analysis is that Nagel's way of looking at theories makes it impossible that two distinct theories should ever disagree or even agree in the claims (or presuppositions) they make about a given set of physical items since: apparent disagreement must be taken as evidence that the claims concern different items; and "agreement" must be taken as evidence that the theories are not really distinct at all.

Unless their common terms are meaning-invariant, one theory cannot be said to deny, without equivocation, a claim or presupposition of another theory. But it is precisely such denials which constitute revolutionary advances in science. Any view which represents these denials as anything less than denials misrepresents, i.e., minimizes, the revolutionary character of such advances. We cannot escape this conclusion by claiming that the theoretical items described by theories T and T_1 need not be distinct if the terms common to T and T_1 differ in meaning, that we need only conclude that the terms are being used differently but that their referents are the same.[11] For if a claim or presupposition of T expressed via terms common to both T and T_1 is a necessary condition for correct application of these terms, then nothing can count as the referents of these terms unless it satisfies this necessary condition. The unpleasant consequence cited above is unavoidable.

Nagel's argument concerning theoretical terms is particularly important because of its evident relation to the arguments of Bohr and Heisenberg. Nagel does not rely solely on this first argument, however. Even if he admitted that quantum mechanics was not deterministic with respect to a "state description" in terms of position and momentum values, he would still maintain that it was deterministic with respect to another state description, that specified by the wave function itself.

Nagel begins his argument concerning the definition of physical state[12] with the claims that a definition of physical state in terms of momentum and position values is not the only definition of state used in physics; that, in fact, this definition is only appropriate in classical mechanics; that hydrodynamics, statistical mechanics, and electrodynamics all use different definitions of state and yet all are rightly considered to be deterministic theories because they satisfy a criterion which is a natural extension of the criterion laid down by Laplace, using classical mechanics as a model:

We ought to regard the present state of the universe as the effect of its antecedent state and as the cause of the state that is to follow. An intelligence knowing all the forces acting in nature at a given instant, as well as the momentary positions of all things in the universe, would be able to comprehend in one single formula the motions of the largest bodies as well as the lightest atoms of the world, provided that its intellect were sufficiently powerful to subject all data to analysis; to it nothing would be uncertain, the future as well as the past would be present to its eyes.[13]

Classical mechanics does determine purely mechanical properties such as positions, momenta, kinetic and potential energies, etc., in the sense that given the specification of the state of a physical system in terms of the positions and velocities of the constituent bodies of the system at any time, the force law, $\mathbf{F} = \dfrac{d\mathbf{p}}{dt}$, enables us to determine uniquely the values of these position and momentum variables at all other times, provided the force functions are correctly specified. Furthermore, the other purely mechanical properties of the system at any time are calculable from known values of positions and momenta at this time. Hence, specification of the physical state of a system (positions and momenta) at any particular time allows, in principle, specification of all mechanical properties at all times.

It is true, as Nagel argues, that other branches of classical physics employ theories—distinct but noncompeting with respect to mechanics—which use definitions of state different from that used in mechanics but which nevertheless are considered by physicists to be deterministic theories. Accordingly, we would like to formulate a general characterization of deterministic theories, a formulation which captures those aspects of these other theories in virtue of which they sufficiently resemble classical mechanics to be regarded as also deterministic. If we can lay down necessary and sufficient conditions for a theory being deterministic, we can then decide, without begging the question, whether quantum mechanics is deterministic, i.e., whether it meets these conditions.

Nagel proposes the following characterization: "a theory is deterministic if, and only if, given the values of its state variables for some initial period, the theory logically determines a unique set of values for those variables for any other period" (1961, p. 292). How are we to apply this criterion? How do we decide what counts as a set of state variables for a given theory? Nagel claims that this decision cannot be made in advance of a "causal" theory concerning the subject matter related to these state variables. That is, he defines a "causal" theory as one containing an equation involving time derivatives and claims that those quantities for which one-one correspondences between earlier and later values are specified by the "causal" equation(s) are the state variables. Thus Nagel arrives at the conclusion that any theory containing equations involving time derivatives is deterministic, provided only that the equations have single-valued time-dependent solutions.

This conclusion is quite suspect, as Nagel himself would admit, since under this criterion it would be very difficult to imagine what an indeterministic theory would be like. All branches of physics use equations with time derivatives and all future physical theories will presumably do so. In spite of its suspect character, we cannot reject Nagel's definition out of hand; rather, we must see how it applies to classical theories other than mechanics, and to quantum mechanics.

In electromagnetic theory, Maxwell's equations constitute a deterministic set of laws with respect to the state variables representing the electric and magnetic field vectors. Given the values of these vectors at every point of an isolated region or regions of space at some particular time, Maxwell's equations uniquely determine the values of the vectors at any other time. In (classical) statistical mechanics, general, nonlinear transport equations constitute a deterministic set of laws with respect to the state variable representing the population distribution of an aggregate of particles as a function of space and energy. Given this population distribution function for an isolated region or regions of space at a particular time, the transport equations uniquely determine this distribution function at all other times.

These two theories, electrodynamics and statistical mechanics, as well as others—e.g., hydrodynamics—are quite different from each other and also from classical mechanics, and yet each satisfies Nagel's criterion and each is considered by physicists to be deterministic. Accordingly, we may safely agree that Nagel's criterion constitutes a necessary condition for a deterministic theory. This is no more than the acknowledgment that unless a theory uniquely relates earlier and later values of some important parameters of physical systems it cannot be considered deterministic. However, the criterion hardly seems to be a sufficient condition. The three cited classical theories have other important characteristics in common which are also necessary for being considered deterministic.

For instance, in each of the classical theories, if one is given a certain degree of detailed information about an isolated system at a particular time, then the same degree of detailed information about all other times is obtainable from the equations of the theory. Another relevant consideration concerning the classical theories is that when a previously isolated system is disturbed there is always, in principle, a way within the theory in question to incorporate the disturbance or perturbation into the system in an exactly correct manner so that the new system—the previously isolated system plus the perturbation—henceforth forms a "deterministic system." The theory in question is, in principle, deterministic with respect to any extension or perturbed state of a previously isolated system.

Let us see how these two crucial points apply to some of the cited classical theories. Consider first the classical mechanics of a single particle. If no external forces are acting on the particle, its velocity will be constant at all times. Given its position at any particular time, its position at all other times and its velocity (constant) at these times is exactly calculable. Thus, so long as the particle may be considered in isolation, the degree of exactness of detailed information theoretically obtainable about the particle is always the same at all times. Now suppose this isolated single-particle system is disturbed by the introduction of a second body, separated from the first by some distance. If this initial separation distance and the initial velocity of the second body are known, then from knowledge of the various mutual forces (gravitational, electrical, etc.) between the two bodies the future positions and velocities of these bodies can be calculated exactly. This conclusion can be generalized to cover any number of bodies and any sort of perturbation.

Two things should be noted here. First, as Laplace implicitly reminded us, although equations representing all such problems, however complex, can always in principle be set up, we may in practice be unable to do so, let alone solve them. Second, the claim that perturbation of previously isolated systems can always be incorporated into the system exactly, applies even if we disturb the originally isolated particle by letting it collide with another. In fact, however, we do *not* attempt to handle such collision problems exactly, even in classical mechanics. Rather, we determine, usually empirically, relative probability of scattering through various angles. In other words, we usually do not know how to specify in detail the impact forces between the two bodies during collision, even for the simplest bodies. But there is nothing in classical mechanics which makes such specification impossible, and, if specified, these impact forces would determine the exact course of individual collisions.

Next, consider a statistical mechanical description of a collection of particles of various energies diffusing in an infinite, source-free, nonabsorbing medium which is at some fixed temperature. So long as this system remains unperturbed, the diffusing particles will be in thermal equilibrium with the (particles of the) medium—the energy distribution of the particles will be independent of space and time. If we now perturb the system by varying the temperature of the medium, the different energy distributions at various times can be calculated exactly from the time-dependent transport equation with the information concerning the time variation of the temperature incorporated. That is, the new, perturbed equation will contain a new energy distribution for the particles of the medium, hence new probabilities for various scattering and/or energy transfer collisions between these particles and the diffusing ones; accordingly, the energy transfer between the two sets of particles will no longer be in balance and the space-energy distribution of diffusing particles will vary with time.

Other examples, from hydrodynamics or electromagnetic theory, could be given, but the above two should suffice.

We are now in a position to see how quantum mechanics differs markedly from classical theories such as those cited above. First,

CHAPTER THREE

consider the matter of isolated systems. It is a fact not sufficiently emphasized in the literature that an *isolated* atom in an excited state can emit radiation, thus giving up energy and undergoing a transition to a less excited state. The point of emphasizing the isolation of the excited system is to note that such transitions can occur even in the absence of any external field (see Sakurai [1967], pp. 36–39). Suppose we have an excited system such that there are several "stable" states with energies less than the excited states. Given the wave function which corresponds to the excited state, the equations of quantum mechanics do not uniquely determine the future states of the system at all times if by 'state' we mean an eigenstate or a physically possible state. What the theory determines are the probabilities that the system will be one particular state or other, probabilities which are single-valued functions of time. So, in general, at any particular time there will be a nonzero probability corresponding to each of several eigenstates. Clearly, in such a situation we cannot claim that the theory has preserved for us the original "degree of exact information" with which we started. We began, by hypothesis, with an isolated system known to be in a particular eigenstate. For later times we do not, in general, know which of its eigenstates the system will be in. We know that it must be in one of the several eigenstates, and we may know that the probability of being in one rather than the other is relatively high. But this theoretical information about the system is simply not of the same exact, unambiguous nature as the original information.

Of course, we can and do represent this indefinite information about later states of the system by a "wave function" which is a linear combination of the functions corresponding to the various eigenstates. It is this new wave function, this linear combination, which Nagel wants us to consider as specifying the "state" of the system. Since the time-dependent Schrödinger equation does determine unique values of this linear combination, Nagel takes this as sufficient grounds for calling the theory deterministic.[14] One suspects that Nagel may have been led astray on this point by the prevailing current usage of physicists, who do indeed refer to the linear combination in question as the "state" of a system, even though at least a few traces of an earlier and stricter usage remain —a usage which reserves the term 'state' for eigenstates. Notice, however, that the issue cannot be settled by superficial appeal to the usage of any particular term. Rather, the issue is whether Nagel's criterion for deterministic theories represents a "natural extension" of that applicable in classical mechanics.

I have argued that any acceptable criterion must demand that deterministic theories preserve the initial degree of detailed information about physical systems. When we find that quantum theory cannot satisfy this demand, we cannot save Nagel's criterion merely by saying that with respect to some parameter the theory contains time-dependent equations with single-valued solutions. Indeed, the example of quantum theory indicates how we can always *construct* such parameters. If the equations of a theory specify one-many rather than one-one correspondences between earlier and later values of some important physical quantity, then we can always form a linear combination of the many later values, according to some unique rule of combination, so that this linear combination has a single value at all times. According to Nagel's criterion, all physical theories sufficiently developed to be worthy of the name are deterministic; there could not be an indeterministic theory.

Under Nagel's criterion, the determinism of quantum theory amounts to no more than saying the theory predicts that an isolated system will always be in one of several physical states at any given time. When the set of physically possible states is discrete, the theory rules out some values for, say, the energy of the system, i.e., it rules out some states. But when the problem at hand involves a continuous spectrum, even this is not the case.[15] Hence, in general, under the criterion in question quantum theory is deterministic in that it predicts physical systems will always be in some state or other and gives the relative probabilities of being in each of the various states. Clearly, this criterion is a far cry from being a natural extension of the criterion relevant to classical mechanics.

Let us now consider two ways of insinuating that perhaps quantum theory does, or at least could, after all, always preserve the original degree of detailed information about a physical system. First, it might be objected that in our spontaneous emission case we should include the emitted radiation as part of the system. Perhaps then the theory will uniquely predict the future states of the (total) system. This suggestion is quite misleading. As we have already noted, the theory predicts only relative probabilities of transition to various states, with corresponding various amounts of energy carried off by the emitted radiation. Of course, the total energy of the combined atom-plus-radiation-field system remains constant; it is the breakdown, the partitioning of energy between the two parts of the system which is not unique. The only sense in which the theory makes a unique prediction is that it "predicts" energy conservation; it predicts that the energy "state" of the total system will remain unchanged. But a process in which an atom makes a transition from E_0 to E_1, emitting $\Delta E = E_0 - E_1$ radiant energy, is quite distinguishable from a process where the transition is to E_2, with $\Delta E_{\mathrm{rad}} = E_0 - E_2$. To defend a deterministic construal of quantum mechanics, we must refuse to admit that these two processes are distinct and must pretend that in fact nothing happens in such processes, since the initial and final "states" are alleged to be the same.

It is amusing to note what happens when we extend this attitude to the universe as a whole: obviously, the total energy of the universe is constant, if we include particle rest energies, so we could claim that quantum mechanics is deterministic, that it uniquely predicts future states of the universe from past states. But the sense of 'state' employed here is the sense in which the "prediction" is possible because nothing ever happens! Clearly, the sense of 'determinism' which can be associated with quantum theory is a Pickwickian sense, a sense no one hankering after classical determinism would be interested in.

A second possible objection is that we cannot specify with certainty which (eigen)state a system is in. Hence, the objection goes, it is misleading to pretend that we "lose" information in going from a quantum mechanical description of an initial state to a description of a later state. That is, it might be alleged that we never had exact information in the first place, hence we did not really lose any; the difference between earlier and later information is not in the information we actually had but only in what we pretended to have. The reasoning behind this objection involves the uncertainty relation between energy and time:

$$\Delta E \Delta t \geqq h$$

This relation imposes a restriction on the accuracy with which the energy of a system can be measured in a finite time. It is natural, though mistaken, to suppose that this relation precludes our knowing which energy eigenstate a system is in at a given (exactly specified) time.

This natural supposition is mistaken because it confuses two senses of uncertainty or dispersion in energy values. Suppose the various excited states $E_1, E_2 \ldots, E_n \ldots$ of the system have mean lifetimes $\tau_1, \tau_2 \ldots, \tau_n \ldots$. Then the mean time available for energy measurement of E_n is τ_n. So

$$\Delta E_n = h/\tau_n.$$

That is, there is dispersion of magnitude ΔE_n in the value of a given energy eigenstate E_n. We cannot know the exact energy value of this eigenstate. But this places no restriction whatever on our knowledge of *which* eigenstate, $E_1 \ldots E_n \ldots$, the system is in at a particular time so long as the energy difference *between* eigenstates is large compared to the dispersion in energy for each eigenstate. This condition is typically satisfied; if it were not, spectroscopic measurements would not yield the familiar discrete line patterns associated with many atoms but rather would yield overlapping band patterns. So we see that the second objection simply does not apply to typical discrete eigenvalue problems that occur in quantum mechanics. It is sufficient for my purposes to have shown that, in general, quantum theory does not preserve initial information. A truly deterministic theory would always do so.

Nagel's criterion fails because it does not incorporate the feature of deterministic theories which I have labeled 'preservation of degree of detailed information.' This consideration alone would be a decisive objection to the criterion, but there is a second and equally decisive objection to be made. We saw that in various classical theories it was always possible, at least in principle, to incorporate descriptions of disturbances of previously isolated systems into the systems in an exactly correct manner. Accordingly, these classical theories were deterministic not only with respect to isolated systems but also to extensions of such systems.

The situation is quite different in quantum mechanics. Consider an isolated system which is initially in its ground state, i.e., its lowest energy eigenstate. To calculate the time behavior of the system when disturbed by external influences, the quantum theorist represents the disturbance or perturbation by a function δH which constitutes an addition to the Hamiltonian function, H_0, containing information about the potential and kinetic energy of the (previously) undisturbed system. The original eigenstate of the system is one of the solutions of the eigenstate problem corresponding to this unperturbed H_0 function. The eigenvalue problem corresponding to the new, total function $H_0 + \delta H$ may or may not have an exact solution, depending on the particular problem, that is, depending on the nature of the physical system and the perturbation.

The important thing to notice is that even in those cases where an exact solution of the new eigenvalue problem is possible, the theory yields only time-dependent probabilities for transition from the original (ground) state to each of the other eigenstates of the system. In general, more than one of these transition probabilities will be nonzero at a given time. Hence when we incorporate information about perturbations of previously isolated systems into our description of these systems the result is not, in general, a "deterministic" set of equations. The set is indeterministic in that it specifies one-many rather than one-one correspondences between earlier and later eigenstates. Quantum theory is not deterministic with respect to extensions of previously isolated systems.

In the next chapter we will consider a "hidden variable" counterargument that attempts to subvert the above conclusion. There it will be shown there is no perturbing function, δH, that will uniquely determine the resulting transition, unless we allow the *functional form* of δH to vary with magnitude of the coordinates of the system. This is like saying that the form of the gravitational force function varies with separation distance—an inverse square law for some distances, an inverse cube for others, etc.

We now turn to consideration of Nagel's argument concerning atomic-macroscopic relationships. Strictly speaking, this is not an argument against the claim that quantum mechanics is an indeterministic theory, but rather an argument to the effect that whatever the status of quantum theory with regard to atomic or subatomic systems we cannot correctly conclude that the theory attributes any "indeterminism" to macroscopic systems. To establish this point, Nagel sets up and demolishes a straw man and never addresses himself to the real basis for the claim that the theory does indeed make this attribution.

The real basis of the claim is that quantum theory applies to bodies of all dimensions, with all physically possible values of momentum, energy, etc. Hence differences between atomic and macroscopic systems—e.g., differences in uncertainty values for important variables such as position, momentum, energy, etc.— are differences of degree—quantitative, numerical differences— rather than differences of kind. Furthermore, and this is a crucial point, the classical description of phenomena, whether atomic or macroscopic, that correctly applies only as a limiting case of the quantum mechanical description is derivable as a special case in quantum theory for physical systems consisting of a single particle. The classical description is derived within quantum theory for systems whose internal structure and dimensions may be neglected. Thus when Nagel, and many others, argue as though the only basis for the claim in question (that quantum theory attributes indeterministic features to macroscopic systems) involves arguments concerning the microscopic composition of macroscopic systems, they indicate that they have completely missed the point. Far from relying on such composition arguments, classical limit derivations assume internal composition to be a negligible consideration. To suppose otherwise is to get the matter completely backwards.

Nagel begins his discussion by setting up the following argument:

Macroscopic objects are complex structures of subatomic ones. The properties and relations of the former therefore occur under conditions that can be formulated in terms of the arrangements and interactions of the latter. But the established theory concerning subatomic objects is statistical and indeterministic: to the best of our knowledge, the behavior of subatomic objects exhibits only statistical regularities. Accordingly . . . since the behavior of macroscopic objects is compounded out of the behavior of their subatomic constituents, the regularities manifested by the former are also only statistical (Nagel 1961, pp. 312–13).

Two things should be noticed about this argument that Nagel sets up and proceeds to demolish. As noted, it fails to mention the real basis for the claim about the indeterminism of macroscopic systems. Second, and equally instructive, the argument can be interpreted, and is so interpreted by Nagel, as making the absurd claim that classical theories concerning macroscopic items are indeterministic in character. But no one who makes the claim in question about macroscopic indeterminism is claiming that any new knowledge, quantum mechanical or other, could possibly perform the feat of changing a theory of one sort (deterministic) into one of a contrary sort (indeterministic). Of course, the nature of particular theories, their deterministic or indeterministic character, cannot change, but our (scientifically based) opinions as to their truth or accuracy can and do change! Those whom Nagel supposes himself to be arguing against are claiming the latter, not the (impossible) former change.

Since Nagel is attacking arguments other than those relevant to the "indeterministic" claims in question, only one further point is necessary. Nagel quite correctly reminds us that "indeterminacy is not exhibited in any experimentally detectable behavior of macroscopic objects" (Nagel 1961, p. 316). This fact is illustrated by the sample calculations presented in chapter 2. But, as Nagel would be the first to admit, the fact that an effect is not experimentally detectable is no indication that a theory does not assert the existence of such an effect. The effects predicted by quantum mechanics differ from those of classical theory throughout the entire range of values of the relevant variables—dimensions, momenta, etc.—and only converge in the limit of infinite values of these quantities; in the range where the differences between predictions of classical and quantum theory are experimentally detectable the latter is in better agreement with experiments than the former. To assume that this superiority of quantum theory over classical theory extends into the range where the differences between the predictions of the two theories are undetectable is merely to make an extrapolative or inductive inference. To suppose that the relative superiority of the two theories is reversed in the range of undetectable differences is to make a counterinductive inference. Anyone who wishes to defend counterinductive inferences as against inductive ones is welcome to do so.

In this chapter I have shown that various arguments against the characterization of quantum mechanics as indeterministic and/or in favor of a deterministic characterization of the theory are unsound. One is simply irrelevant; one presupposes untenable views concerning the nature of theories and theoretical terms; one trivializes the notion of determinism beyond recognition. We may conclude that quantum mechanics—and, by extension, those physical systems to which it applies, namely, all systems—is indeterministic. Whether this indeterminism is here to stay, only time will tell—that is, only future developments in physical theory will tell. We should be somewhat hesitant in saying what the world is "really" like. But at least we can and should get clear about what our latest and best established theories are like.

The question of the determinism or indeterminism of quantum theory is intrinsically important, but it is also important because of its relation to other aspects of the interpretation of quantum mechanics. In the next two chapters I discuss two moves which have been offered as alternatives to the standard Copenhagen interpretation, the first involving the introduction of hidden variables and the second a proposed revision of logic. In both cases, the alternatives seem inspired by an unjustified determination to have quantum mechanics be deterministic.

4
Hidden Variable Theories

There are two points to be made concerning the hidden variable "interpretations" of quantum mechanics that have been offered by Bohm, de Broglie, and many others. First, these hidden variable theories are just that—new theories, and not new interpretations of quantum mechanics. These new theories are consistent with the empirical evidence they were designed to explain, but they are not supported by this evidence. Second, these theories are conceptually consistent alternatives to the standard quantum theory, alternatives not to be rejected out of hand, formulations and investigations not to be put on a par with the design of perpetual motion machines. I emphasize both points because most authors emphasize one while ignoring the other, thus presenting a misleading picture. Examination of Bohm's theory will reveal that the motivation behind hidden variable theories is the restoration of determinism and that such a restoration is misguided, i.e., unjustified on present evidence.

Bohm suggests two quite distinct alternatives to the standard quantum theory. First, he claims that even if we retain exactly the same mathematical formalism as in the standard theory we can still interpret the equations so as to obtain a deterministic explanation of all quantum mechanical phenomena, an explanation employing particles and fields, where 'particle' and 'field' are to be understood as in classical physics.[1] On this first procedure Bohm's theory yields the same experimental predictions as the standard theory; no direct experimental investigation would enable us to choose between the two theories. Second, Bohm suggests ways in which the equations of the standard theory could be altered so that, although both the standard theory and the altered theory yield the same predictions for physical systems of linear dimension larger than (approximately) 10^{-13}cm, the predictions of the two theories diverge radically in the range below 10^{-13}cm. Accordingly, in principle it would be possible to design experiments which would enable us to choose between the two theories, though we are not ready to perform such experiments.

The central idea of Bohm's first alternative is that, in addition to forces due to classical potentials, particles are subject to forces due to a "quantum-mechanical" potential. It will be recalled that the classical equation of motion for a single particle is obtained by assuming the negligibility of a ψ term proportional to h^2. Bohm suggests that we retain this term, interpret it as a potential, over and above the classical potential accounted for in a separate term, and then interpret the equation as a quasi-classical equation of motion, uniquely and exactly describing the individual paths of individual particles. The source of this added potential Bohm takes to be a field whose strength is related to the amplitude of the Schrödinger wave function. Thus, "just as the electromagnetic field obeys Maxwell's equations, the ψ-field obeys Schrödinger's equation" (Bohm 1952, p. 170).

In applying this new theory to standard problems, Bohm arrives at some strange results. In calculating the steady state solution of the Schrödinger equation for particles with zero angular momentum, he finds that the particles are always motionless. This result would seem to be in conflict with experiment, but Bohm claims that here and, indeed, in all experimental situations, the apparent conflict disappears if we extend his new "interpretation" so as to include the measuring apparatus itself. That is, Bohm claims that in general the results of measurements are in principle determined by the initial form of the wave function for the combined system—apparatus plus observed item—and by the initial coordinates of the observed system and of the apparatus. In practice these initial coordinates can neither be predicted nor controlled, and, since Bohm's quantum-mechanical potential is extremely sensitive to the values of these coordinates, the results of the measurements are not uniquely predictable. Rather, all that is predictable are the results of a statistical ensemble of measurements—just as in the standard theory.

Thus, in the zero angular momentum steady-state case we measure a spectrum of momentum (or velocity) values even though the particle is standing still before the measurements are made, and, in principle, the interaction between the particle and the measuring apparatus uniquely determines the result of each individual measurement. Similarly, the transitions between stationary states induced by interactions with other systems are asserted to be uniquely determined in principle by the wave function representing the combined system—original system plus perturbation—and by the initial coordinates of the two systems. But, once again, in practice these initial coordinates are not precisely known, not predictable or controllable, hence we have recourse to statistical predictions for the transitions between stationary states.

The results of Bohm's analysis of the well-known two-slit diffraction experiments are stranger still. Here the problem is to explain why none of the particles that pass through a double slit are able to reach certain points on the detecting screen, points which some of them do reach if only a single slit is open. How can the mere opening of a second slit prevent these particles from reaching these previously reachable points? Bohm's explanation is that the quantum mechanical potential changes radically when the second slit is opened. The force on a given particle becomes infinite as the particle approaches one of the "forbidden" regions on the detector screen. If the approach is from a certain direction, the force is a repulsive one, preventing the particle from reaching the region. If the approach is from the opposite direction, the force is attractive rather than repulsive, but since the force is infinite the particle acquires an infinite velocity and literally spends no time in the region where this infinite attractive force exists.

What are we to make of these strange results and the new "interpretation" which requires them? In particular, what is the point of my earlier insistence that this new alternative be regarded as a new theory rather than a new interpretation of quantum mechanics?

In a sense there really is no distinction between theories and interpretations of theories; theories cannot be identified with their formalisms, logical and/or mathematical. It is misleading to talk about interpreting a theory since, strictly speaking, a formalism is not a theory at all until it is interpreted. In particular, the formalism of a physical theory is not to be regarded as a theory until an interpretation, of some minimum degree of definiteness, in terms of physical items—objects, structures, processes, etc.—has been given. It is dangerous to speak of "alternative interpretations of a theory" and thus conjure up an image of a theory which exists independently of any and all interpretations of "it" and which remains unchanged in going from one interpretation to another, radically different interpretation.

Of course, we can reject this view of theories and still admit that *minor* modifications in the formalism and/or the physical interpretation of a theory need not compel us to regard the modified version of the theory as a distinct new theory. (For example, it does not seem that the general theory of statistical mechanics, a theory based on nonlinear transport equations, is a different theory from the special theory based on linear equations.) But the alternative proposed by Bohm to the standard quantum theory is nothing like a mild, minor modification. It is a drastic, radical reinterpretation, postulating a new, hitherto undetected field which the standard theory presumes does not exist. Bohm's theory is a quasi-classical, deterministic theory in which the individual paths of individual particles are in principle calculable, whereas the standard theory is inherently indeterministic. Bohm's alternative must be regarded as a distinct new theory.

Bohm's new theory is not supported by the observational consequences it was designed to explain. Our concept of empirical support involves much more than successful prediction. Even if two theories or hypotheses always yield the same predictions concerning observations, we can and do judge the relative probabilities of the two theories being true on the basis of several other criteria, including simplicity, initial plausibility, and (allegedly) extendibility of prediction. Before applying these criteria to Bohm's ideas, we must answer Heisenberg, who specifically denies that two theories with the same set of observational consequences are even distinguishable as different theories:

we are thus concerned not with counter-proposals to the Copenhagen interpretation but with its exact repetition in another language. . .the hidden parameters of Bohm's [first alternative] interpretation are of such a kind that they can never occur in the description of real processes. . . . Bohm's language. . .says nothing about physics which is different from what the Copenhagen language says (Pauli 1955, pp. 17–19).

Heisenberg is here presupposing that a theory is to be identified with, entirely characterized by, the sum total of its empirical or observational predictions. If no observation could ever enable us to decide between two theories, then—Heisenberg claims—the two theories are not really distinct at all even though they may be couched in language which suggests important differences. The claim presupposed by Heisenberg is false but it is easy to understand its initial attraction. If the two theories in question really asserted different claims about physical structures, processes, etc., then we would expect that eventually the differences would show up—that there would be some way of detecting, however indirectly, the asserted differences. Conversely, if by hypothesis such detection is impossible, then we strongly suspect that no differences are really being asserted. Notice that one need not be an empiricist or positivist to appreciate the intuitive attraction of Heisenberg's claim. We can hold a realistic interpretation with regard to theoretical terms corresponding to items not directly observable and still feel that, unless the observational consequences of two sets of claims made using these terms are at least slightly different, then the two sets of claims are not really different. In fact in the normal situation—where the two sets of observational consequences are not fixed as the same by hypothesis—it is just because we interpret the theoretical terms in a realist fashion that we expect the different behavior of the items corresponding to these terms to produce different observational consequences, consequences causally connected to the particular structures or processes described by the theories. But the normal situation does not correspond to the case under discussion.

The falsity of Heisenberg's claim is apparent once we perceive exactly what lends it plausibility. The force of the claim depends on the tacit assumption that, if the two sets of observational consequences are the same, then we have no way of deciding between the two theories, no more reason to believe one than the other, no justification for saying one is better supported by the (observational) evidence than the other. This tacit assumption represents a non sequitur. There are other, more or less independent, criteria, over and above mere consistency of predictions with observation, for evaluating the relative degree of empirical support for theories and/or hypotheses; if we can show that two theories differ in their relative empirical support, this is sufficient to show that they are distinct theories, that they are not identical. One obvious consideration is simplicity, another is initial plausibility or probability, a third is (allegedly) extendibility of predictive power.[2] As for simplicity, the point to remember is that all other things being equal— and unfortunately they seldom are—the simplest relevant hypothesis consistent with the observed facts is the hypothesis best supported by these facts.

This principle of simplicity or economy is frequently misunderstood. It is not intended by its advocates, certainly not by this writer, as either an aesthetic or a pragmatic principle. When we claim that a given hypothesis or theory is better than a competitor because it is simpler, we are not claiming that it is better because it is somehow more elegant, more pleasing to some sense of intellectual beauty. The best theories are not the prettiest but the truest. Similarly, we are not claiming that the better theory is pragmatically superior in that it is easier to use efficiently; this claim, though usually true, is irrelevant, since truth, not efficiency in finding the truth, is the item at issue. In comparing the relative degree of support for competing theories or hypotheses, we are not asking about the most efficient means of discovering the truth about some subject matter; we are asking which theory most nearly approximates the truth insofar as we are able to decide on present evidence. So the principle under discussion is not a pragmatic or an aesthetic one and certainly not an arbitrary, stipulative one; rather the principle is descriptive, asserting a conceptual or analytic connection between our notions of simplicity on the one hand and degree of empirical support on the other.[3]

We cannot pretend that this notion of simplicity is a straightforward and untroublesome one. Broadly speaking, there are two kinds of simplicity relevant to confirmation—descriptive simplicity and ontological simplicity. Furthermore, descriptive simplicity is ambiguous in that it may seem to refer to the complexity of the equations of a theory, the complexity of their solutions, the number and/or variety of theoretical assumptions, or some combination of these aspects. Ontological simplicity refers to the number and variety of items whose existence is postulated or presupposed by a theory. These distinctions make the application of the principle of economy very difficult in many cases, since a given theory may be simpler than a competitor in some respects but less simple in other respects.[4] Nevertheless, the truth of the principle can be brought out by examining its application to clear-cut cases.

Consider first the oft-cited standard case of curve-fitting a set of values of one variable (y) as a function of another variable (x). Given a discrete, finite set of (x,y) values, what hypothesis concerning the general relationship between y and x—i.e., what function $y = y(x)$—is best supported by the evidence which the discrete, finite set of (x,y) pairs represents? Which function is most likely to provide the most accurate interpolation between known (x,y) pairs and extrapolation beyond the range of these known pairs? Assume for the sake of definiteness and convenience that the functions $y(x)$ are restricted to continuous, single-valued functions. In this case, the "best" choice for a fitting function which includes all known (x,y) pairs is that function with the minimum number of turning points per x-interval. To see why this is so, consider the particularly simple case where all the known (x,y) pairs can be included, to a very good approximation, on a straight line. The simple straight-line hypothesis, $y = a + bx$, is better supported by the evidence than any of its more complex competitors, even those many competitors which can be chosen so as to include all the so-far known (x,y) data points. To see why this is so, consider, for the sake of definiteness, a particular oscillatory function which includes all the known (x,y) pairs but which has y-values radically different from the straight line values at x-values in between the known x-values.

One way of convincing ourselves of the stronger empirical support attaching to the linear as compared to the oscillatory hypothesis is to notice that if the oscillatory hypothesis were true we would have expected to obtain some (x,y) points different from the linear values, provided only that the data were not taken in a completely biased way. As we take more and more steps to insure randomness of data points, this expectation is enhanced. These facts are not enough to establish the real point of interest. For we can suppose that an opponent modifies the form of the oscillatory function so that its amplitude or deviation from the straight-line values is very small and, further, claims that there is some limit, whether practical or theoretical, to the precision with which the (x,y)

measurements can be made, a limit such that the differences between the linear hypothesis and the modified oscillatory hypothesis cannot be observationally detected. In this modified situation there would be no difference between the "expectations" corresponding to the two hypotheses; we would expect the straight-line data no matter which hypothesis were true.

So, in general, it is not so much a matter of our expectations as a matter of what counts as confirmation of our expectations. If we believe the linear hypothesis is true, we expect linear data; if obtained, these data count as confirmation of our expectations. Even if we had no expectations one way or the other, the linear data, when obtained, support the linear hypothesis precisely because the data confirm the expectations we would have had if we previously believed the hypothesis to be true. The situation is quite different with respect to an oscillatory hypothesis. Here we expect oscillatory data if we believe the hypothesis is true, and the hypothesis is confirmed if and only if such data are obtained. In the case where the oscillatory amplitude is alleged to be too small to be detected, we can admit that the linear data do not falsify the oscillatory hypothesis, although the important point is that the data do not confirm the hypothesis either. The move from "not falsified" to "confirmed" is based on an equivalence whose falsity is evident once the equivalence is made explicit. The notion of confirmation or positive support involves far more than mere absence of falsification. The mere fact that a hypothesis has not been falsified does not in itself indicate that the hypothesis has received the slightest confirmation.

The important point to keep in mind is that what counts as confirmation of a hypothesis is data of complexity comparable to the hypothesis itself. This is obvious in cases such as the one under discussion where the predicted data points are simply instances of general hypotheses. A linear array of data points is just a confirming set of instances of the linear hypothesis. Nothing but an oscillatory set of data points would count as a set of confirming instances of the oscillatory hypothesis. Of course, *if* we had specific evidence from which to infer that suitably corrected data would be of oscillatory form in spite of the linear form of the raw data, *this evidence* would constitute support, of an indirect sort, for the oscillatory hypothesis. But it is one thing to have such specific positive evidence, it is quite another thing merely to hypothesize reasons why the data appear linear rather than oscillatory. In the first case, we could claim to have indirect support for the oscillatory hypothesis; in the second, we can only claim that the data do not conclusively falsify the oscillatory hypothesis. Sooner or later the proponent of the more complex hypothesis must provide some specific positive evidence. He cannot claim "support" or "confirmation" by continually introducing new hypotheses to explain why his earlier ones do not seem to be confirmed. Hypotheses are not substitutes for evidence.

It is precisely at this juncture that we can perceive the source of Heisenberg's error. Ordinarily we would expect, eventually, to be able to detect, directly or indirectly, the differences asserted by two competing hypotheses or theories. But we can always change our expectations for one theory by adding additional hypotheses which, if true, would make such detection impossible. The complex theory embodying such ad hoc hypotheses does not make the same assertions about the physical world as its simpler competitor, nor are its assertions meaningless—i.e., neither true nor false; rather they are assertions which have received not the slightest positive confirmation. Hence, if its simpler competitor is confirmed by the observed evidence, we have good reason to believe the complex theory false and, if false, not meaningless and certainly not indistinguishable from its simpler competitor.

Points analogous to those made above can be brought out in cases involving ontological simplicity. Consider the well-known example of Reichenbach (1946, pp. 18–20), involving our perceptual experience of familiar objects such as houses, trees, etc. By far the simplest hypothesis which accounts for the regularity and (partial) continuity of our perceptual experience is the hypothesis that such familiar objects continue to exist when not being observed. But other, more complicated hypotheses are also consistent with the perceptual evidence. For example, we could hypothesize that the familiar objects of perception cease to exist when we stop observing them and come into existence again when we resume observation. Why then do we prefer the more natural, common-sense hypothesis of continued, independent existence of external perceptual objects, to the fantastic hypothesis of a neon-world? Contrary to what Reichenbach claims, this preference is surely not a "conventional" one. Rather, the justification of our preference for the simpler over the more complex hypothesis is exactly analogous to that given in the curve-fitting case.[5] Our perceptual experience is consistent with both hypotheses, it falsifies neither of them; but it confirms the simpler hypothesis and does not in the least confirm the complicated one. Once again, the point is that what counts as confirmation of a hypothesis is evidence of complexity comparable to that of the hypothesis itself. We cannot avoid passing unfavorable judgments on the relative degree of confirmation of more complex hypotheses by bringing in ad hoc assumptions which explain why the complex hypothesis has not been falsified. To do so betrays a Popperian confusion resting on a false equivalence between mere lack of falsification and positive confirmation, an equivalence ultimately no better than the traditional material fallacy of "appeal to ignorance."

Given the acceptability of the principle of simplicity in clear-cut cases, we can apply the principle to Bohm's first suggestion. It is clear that Bohm's theory is as complex or more complex than the standard theory in all relevant senses of complexity, whether descriptive or ontological. Since Bohm is taking over the equations of the standard theory unchanged, the two theories are equally complex on this score. However, Bohm intends these equations to be solvable in principle for the individual paths of individual particles; but because of the sensitivity of his quantum mechanical potential these solutions would be—if he actually obtained them, which he never does—extraordinarily complicated. Hence, on this score his theory is more complex than the standard theory. More important, Bohm's theory is more complex ontologically in that he postulates the existence of a hitherto undetected and, by hypothesis, undetectable field—the field responsible for his quantum mechanical potential. Precisely because of these complexities Bohm's suggestion must be regarded as most improbable compared to a more standard, indeterministic interpretation. When we find that Bohm's predictions regarding the (zero angular momentum) steady-state case or regarding transitions between stationary states are not confirmed by the observational evidence, we cannot save his theory by introducing further assumptions to explain the lack of falsification of the theory.

Once again the same conceptual points regarding what counts as confirmation apply here as in the perception case and the curve-fitting case. Similarly, the sensitivity of Bohm's new field, which accounts for its undetectability and hence for the lack of falsification of the claim of its existence, cannot count in favor of the claim of positive confirmation. Bohm himself indicates that he would accept these judgments concerning the lack of confirmation of this theory compared to the standard theory, judgments based on considerations of simplicity. He writes that "of course, we must avoid postulating a new element, i.e., physical object, structure, process, etc. for each new phenomenon" (1952, p. 189), presumably because he accepts the conceptual points cited above. But Bohm implies, quite properly, that simplicity is not the only relevant criterion and argues that, when all the criteria are taken into account, his theory is on the whole epistemologically superior to, and more likely true than, the standard theory.

Immediately after his admonition against proliferating new entities, Bohm claims that

an equally serious mistake is to admit into the theory only those elements which can now be observed. For the purpose of a theory is not only to correlate the results of observations we already know how to make, but also to suggest the need for new kinds of observations and to predict their results. In fact, the better a theory is able to suggest the need for new kinds of observations and to predict their results correctly, the more confidence we have that this theory is likely to be a good representation of the actual properties of matter and not simply an empirical system especially chosen in such a way as to correlate a group of already known facts (1952, p. 189).

The only issue for us here is whether Bohm's theory is better supported by the known evidence than the standard theory because his is capable of suggesting new (kinds of) observation and predicting the results of these new observations (earlier discussions are sufficient basis for rejecting Bohm's insinuations that anyone who disagrees with him is a positivist).

Remember that we are endeavoring to decide which theory is better supported on present evidence. If the *new* predictions of a hidden variable theory turned out to be correct, then that theory would, all other things being equal, be better supported, since it would be supported by a wider and more varied range of evidence. On the other hand, the new predictions could turn out to be false. In any case, it is idle to speculate about what new evidence might be obtained. All we can usefully do is decide if the mere fact that a theory makes new predictions lends support to that theory.

Bohm's attitude on this question is ambivalent. At times he seems to confine himself to the innocent claim that we ought not completely ignore hidden variable theories since it is certainly possible that they might prove correct in the end; at other times he seems to be saying hidden variable theories should already be regarded as superior to the standard indeterministic theory, completely apart from eventual experimental evidence, just because they are more suggestive of new experiments. It is possible that Bohm and I are arguing at cross-purposes here. Perhaps Bohm would admit that hidden variable theories are not more likely to be correct but would still claim they are "superior" in some other sense—e.g., they maximize predictability or testability. If *this* is Bohm's position, it is rather odd. Of course, we want our theories to be testable, to make lots of predictions. But we do not make predictions and test them because such exercises are ends in themselves! We test theories to find out which ones are correct, and which incorrect.

Bohm's *first* suggestion—in which the equations of the standard theory are taken over unaltered—is self-defeating if judged by the cited criterion of extendibility of predictive power. For, by hypothesis, the predictions of Bohm's theory are exactly the same as those of the standard theory; Bohm's theory could not possibly lead to new kinds of observations that would not also be suggested by the standard theory. The altered equations of Bohm's *second* theory do lead to new (kinds of) predictions, predictions which in principle though not in practice enable us to decide between Bohm's second theory and the standard quantum theory. But we have no reason to consider Bohm's second theory more likely to be true than the standard theory. On present evidence, which, after all, is the only thing present judgments can be based on, we should not conclude that the altered equations are better supported because they suggest observations we would not consider making if we accepted the standard equations.

It is helpful here to invoke the examples cited above. Consider first the curve-fitting example; in particular, the version wherein the limitation on the precision of measurements is a practical rather than a theoretical one. Is the oscillatory hypothesis, in this case, better supported by the (linear) data than the linear hypothesis merely because if we believed the former hypothesis we would be led to improve observational techniques in the hope of detecting the small-amplitude oscillations, while if we believed the latter we would be content with the present evidential situation? Clearly, the answer is no. The burden of proof is on the proponent of the oscillatory hypothesis. If more refined measurements yield oscillatory data then, but only then, his hypothesis is vindicated. On the basis of the old linear data, there is no reason to believe that the hypothesis is true.

What lends even the slightest superficial plausibility to the "criterion" of predictive extension is simply a confusion between falsifiability and confirmation; it is true that the purpose of theory construction is better served, all other things being equal, by a theory which leads to and correctly predicts new (kinds of) observations. But this is only to say that more powerful, more general, more fruitful and suggestive theories are better established than more restricted, limited, unfruitful and unsuggestive ones. The above adjectives, however, are retroactive compliments bestowed on theories or hypotheses after they have received varied and extensive confirmation, after we have attempted to falsify them and failed. The mere fact that they are more falsifiable than their more restricted competitors provides no a priori grounds for judging them more likely to be true before they are put to the test. And the criterion of simplicity reminds us that we have good grounds for believing their competitors better supported in the absence of new evidence. When Bohm claims hidden variable theories are superior to the standard indeterministic quantum theory, he illicitly moves from the claim that hidden variable theories are more falsifiable to the claim that they are better confirmed—more likely to be true— a move which would only be justified if, contrary to fact, we have already attempted to falsify them *by new observations* and failed.

So far we have seen that Bohm's first theory and, a fortiori, his second theory are most improbable compared to the standard indeterministic theory according to the criterion of simplicity, and that this judgment cannot be reversed by appeal to a "criterion" of predictive extension. There is a third relevant criterion, initial plausibility or probability, and a defender might well claim that Bohm's theories are more likely to be true because, when all is said and done, they are more initially plausible, relating this claim to Bohm's restoration of determinism. That is, a defender might claim that the past success of deterministic theories is sufficient to support the conclusion that a deterministic theory of quantum mechanical phenomena is more likely to be true than the standard indeterministic theory.

Bohm himself advances a specific argument along these lines. He reminds us that whenever in the past we have had recourse to statistical theories we have always eventually discovered that the individual members of the statistical ensembles obey deterministic laws, laws expressed in terms of previously "hidden" variables. Accordingly, a defender of hidden variable theories might argue: "Every wide-ranging, well-established physical theory developed in the past has either been a deterministic theory or has eventually been superceded by a deterministic theory covering—i.e., predicting, explaining, describing, etc.—the same phenomena. Hence we have good inductive grounds for supposing that the standard indeterministic quantum theory will eventually be replaced by a deterministic theory covering the same phenomena." This argument, in spite of its superficial attractiveness, is unacceptable because it involves a fallacy which amounts to performing the induction on the wrong variable or, better, to ignoring part of the total evidence. We cannot simply rely on induction by enumeration, where the number involved is the number of past physical theories. The relevant variables to consider in assessing the relative likelihood of deterministic vs. indeterministic theories are quantities related to the physical systems described by theories—quantities such as momenta, energies, dimensions, etc. The deterministic theory (classical mechanics) which "superceded," in Bohm's sense, previous statistical theories is, strictly speaking, incorrect for all physical systems, for systems ranging over all momenta and dimension values, while the indeterministic quantum theory is, on present evidence, correct for all physical systems.

Once we concentrate on the proper variables we see two things. First, an extrapolative or inductive argument supports the conclusion that all physical systems are indeterministic, not the opposite conclusion that they are deterministic. Second, the same evidence which yields this conclusion also explains why the deterministic theory (classical mechanics) was able to "supercede" previous statistical theories—because the differences between the predictions of deterministic classical mechanics and those of indeterministic quantum mechanics were experimentally indistinguishable until quite recently. Now that new experimental evidence is available, we cannot ignore this evidence in assessing the relative likelihood of the two types of theory, deterministic or indeterministic; and it is only by ignoring this evidence that the inductive inference based on the past success of deterministic theories can be made to seem plausible.

Even if a defender of hidden variable theories were convinced of the fallacy of the argument from the past success of determinism, he would still very likely be uneasy about the prospect of abandoning determinism and embracing indeterminism. On several occasions Bohm characterizes his deterministic hidden variable theory as "rational" and contrasts it with the standard indeterministic theory which is held to be irrational or involving postulation of processes "not subject to. . .rational analysis." Given this loaded description of the two theories, it is easy to see why someone might be uneasy about accepting the standard theory and rejecting hidden variable theories. After all, who would choose an "irrational" theory over a "rational" one? But this way of posing the question is easily shown to be confused and illegitimate. Specifically, there are at least two confusions involved here. First, the concepts of determinism and causality are conflated; no room is allowed for any distinction between the two. Second, the "reasons" (i.e., causes) or absence of reasons (causes) for the occurrence of various processes are identified with the rationality or irrationality of believing the theory postulating such processes.[6]

Waiving the second confusion, consider the claim that the theory in question does postulate causeless events and it is part of our concept of causality that there are no such events; hence the theory cannot really be satisfactory and we must presume that it will eventually be replaced by a theory in which the causes of these events are specified or at least in principle specifiable. Such an argument carries no weight because the standard quantum theory does not assert or presuppose the occurrence of causeless events. Recall that events such as transitions between eigenstates of a physical system are in general caused by perturbations or disturbances due to external influences, disturbances quite capable of causing the transitions—"capable" in the sense of introducing sufficient energy into the formerly isolated system to account for the transition. Furthermore, even in those special cases where isolated physical systems undergo transitions in the absence of external disturbances, the transitions are causal events in the sense that the earlier states of such systems contained sufficient *internal* energy, momentum, etc., to account for the later (i.e., after transition) values of these quantities.

These points would not satisfy anyone who insists on identifying causality with determinism. That is, an opponent might insist that unless similar causes—earlier, pretransition, states indistinguishable from one another—produce similar results—transitions to the same later state—then we have no right to suppose we have really specified the actual, precise cause of any particular transition which does in fact occur. In other words, unless these events are deterministic —with a one-one correspondence between earlier and later states— they are acausal events. But we really cannot identify causality with determinism. Consider the following examples. First, suppose that we suddenly introduce an object such as a human being into an environment consisting of a near-perfect vacuum—e.g., outer space. The object will explode, because its strong internal pressure will no longer be balanced by the usual external pressure of, in this case, air. Suppose we do this many times, with objects as nearly indistinguishable as we can make them, and observe the distribution of trajectories of explosion fragments. If it is found that a given set of initial conditions does not lead to a unique distribution of trajectories, should we conclude that these trajectories are causeless events? Surely not. The cause is the same in all the explosions—the pressure imbalance between the inside and outside of the object.

The objection will be made that each explosion has a slightly different cause even if we cannot distinguish these different causes; if we could specify the initial conditions exactly, there would be a one-one correspondence between earlier and later conditions. But what lies behind this claim? Nothing more than a faith that some deterministic theory—e.g., classical mechanics or something generically similar—is true.[7] Why should we believe this? On present evidence we must conclude that no such theory is correct. If the claim that all events are deterministic is an empirical claim, all the evidence is against it. If it is to be defended, it must be defended on a priori or conceptual grounds. That is, a defender must rely on the "fact" that we do not count an earlier state of affairs as the cause of a later state unless there is a one-one correspondence between such earlier and later states. But this putative fact is not a fact at all. We do label certain initial states or conditions the causes of later states without attempting to decide if conditions indistinguishable from these particular initial conditions would always yield the same result.

In the explosion case described above we unhesitatingly attribute the cause of the explosion to the pressure differential. Similarly, if a speeding car collides with a pedestrian and hurls him through the air, we unhesitatingly assert that the cause of the victim's unfortunate trajectory was the impact delivered to him by the car. If we found that cars with momenta, energy, etc., indistinguishable from the particular values in question did not produce exactly similar trajectories, would we then withdraw our claim that the car's impact was the cause of the original trajectory? Surely not; no more than we would in the explosion case. In such cases what we mean by saying that certain initial states are the causes of later states is that the earlier states were capable—contained sufficient momentum, energy, etc.—of causing the later states and did in fact produce them. Whether the earlier states produce the later ones in deterministic fashion—whether the correspondence between earlier and later states is one-one rather than one-many—is a separate question; frequently we do not even raise this question, let alone wait upon its answer, in assigning causes.

Hence the identification of causality with determinism cannot be said to be based on a description or analysis of our actual use of causal language. And if it is intended as a prescription or proposal for revision of such language, there is nothing to be said in favor of it—it is based on the presupposition that determinism is true, a presupposition that present evidence denies; and of course if the identification of causality with determinism is prescriptive, not descriptive, the argument of the hidden variable theorist becomes a very curious one. The argument then amounts to saying that indeterministic theories are unacceptable because they are not the sort of theories which would be true if determinism were true—if the structure of physical events corresponded to the way we "should" use causal language, rather than to the way we do use such language. Whatever reasons one can advance for a proposed revision, there is one reason which must be considered rather a poor one: we should not revise our use of causal language merely to make it conform to a belief whose denial is supported by present evidence.

Parenthetically, we should note that the motive for the spurious identification of causality with determinism seems to be traceable to a confusion between the strict universality alleged to be involved in causal claims and the generality—in the sense of generic similarity—actually involved. Clearly, we do not label earlier states or events as "causes" of later states or events unless the earlier states or events always produce effects which are generically similar. If the indistinguishable high pressure differentials did not always lead to explosions, with the same total energy and momentum carried off by, though differently partitioned among, the several fragments, we would not label the initial pressure differential the cause of each of these resulting distributions of fragment trajectories. Similarly, if the speeding car did not always impart some large momentum to the pedestrian, we would not call the car's impact the cause of each of the resulting trajectories. But this is only to say that causes produce generically similar results, results all consistent with conservation laws. If this were not so, we would hardly be able to notice enough similarity in physical processes to recognize and isolate causal sequences. However, it would be completely unwarranted to move from this truistic claim to the false claim that the effects of indistinguishable causes must be identical. Generic similarity does not entail identity, and generic similarity of effects is all that is reflected in our actual use of causal language.

Hidden variable theories must be regarded as bereft of positive empirical support or confirmation, as judged by either of the major criteria of empirical support—initial plausibility or simplicity. The bulk of this chapter is devoted to criticisms, but it is important that we do not move from the conclusion that hidden variable theories possess no positive empirical support to the stronger, and unjustified, conclusion that such theories are fundamentally illegitimate, that their formulation and investigation are somehow radically misguided enterprises. Usually such claims are alleged to be based on a famous theorem of von Neumann. The details of the proof of this theorem are lengthy and abstruse, but for present purposes the following summary description will suffice.[8] Von Neumann begins by assuming that physical quantities are representable by operators having various mathematical properties, notably linearity; he further assumes the usual quantum mechanical rules for calculating probabilities. That is, in effect he assumes that physical systems are correctly described by the Schrödinger wave equation or its mathematical equivalents and assumes the usual probabilistic or statistical interpretation for the wave function which is the solution to this equation. Von Neumann then notes that all physical systems display some dispersion: at least some of the quantities representing physical properties of these systems will have some indefiniteness or unreproducibility associated with their values. This minimal dispersion is predicted by the standard quantum theory and experimentally confirmed. Now suppose we assume that items corresponding to "hidden" variables are operative in such systems, the values of the hidden variables being dispersionless and the values of the observed quantities, some of which display dispersion, *all* being exactly determined by the values of the hidden parameters. The theorem of von Neumann shows that this additional assumption is inconsistent with the previous assumptions—i.e., it is inconsistent with the statistically interpreted Schrödinger wave equation.

So von Neumann has proved that so long as quantum theory retains its present mathematical form and so long as no evidence is found which falsifies this well-established theory, no hidden variable theory could sensibly be regarded as a correct interpretation of the evidence. It is clear from the above description that those who claim, on the basis of von Neumann's elegant theorem, that *no* hidden variable theory could be correct are simply begging the question. For, as Bohm quite properly points out, the theorem is irrelevant to the situation where a theory of the present mathematical form is a good approximation for the range of values of physical quantities presently measurable but is not (exactly) correct outside this range. A new theory, accurate for a wider range, might be of such a mathematical form as to circumvent von Neumann's conclusion by failing to satisfy his premises. In particular, the equations of the new theory might not display the requisite linearity. So long as extensions in the range and kind of measurement are possible, and there seems no reason to doubt that this means forever, the formulation and investigation of hidden variable theories is a legitimate enterprise.

Bohm is not content with noting that von Neumann's theorem is irrelevant to theories of mathematical form different from that of the standard theory. He also claims that the conclusions of the theorem can be circumvented even for theories with the same mathematical form—i.e., the same equations—as the standard theory but with a (radically) different physical interpretation of these equations. That is, von Neumann presumed that the observable quantities displaying dispersion are properties of the observed system. Bohm wants to identify these observables displaying dispersion with properties of the combined system (observed system plus apparatus) and claims that the observed dispersion can be explained away as a distribution over values of hidden parameters operative in both the observed system and the apparatus.

It is difficult to make a straightforward evaluation of Bohm's claim, to connect his discussion with that of von Neumann, simply because Bohm never actually carries out his new interpretation with respect to measurement apparatus but relies instead on qualitative arguments. Nevertheless, there is a way we can sort out and evaluate the claims and counterclaims involved in this issue. Remember that the interactions involved in measurement are a subclass of physical interactions in general. We need not examine Bohm's (nonexistent) application of his new interpretation to measurement interactions; we can evaluate his interpretation with regard to some other kind of physical interaction. It will be convenient to discuss the case of transitions between stationary states of a given system, where these transitions are induced by interaction of the system with an external system.

Bohm chooses to discuss inelastic scattering cases as examples of such externally induced transitions, with free electrons as incident particles and atomic systems with bound electrons as targets. In an inelastic scattering event, the incident electron exchanges part of its energy with the target atom, this energy being used to induce a transition from one bound state, of lower energy, to another, of higher energy. Bohm claims that for such inelastic scattering events we cannot predict the individual outcomes—i.e., we cannot predict which of several possible transitions will occur; but this inability is due, according to Bohm, entirely to our ignorance of the initial form of the wave function representing the combined, interacting systems. He insists that if we knew these things we could, in principle, predict these individual transitions rather than merely predicting their relative probability. This claim of Bohm's sounds plausible, but there is something very suspect and misleading about it. Bohm never actually carries out a sample calculation where he traces the motions of both systems throughout the interaction. This would not be feasible because of the complexity of the trajectories in question, this in turn being due to the sensitivity of the quantum-mechanical potential to the values of the relevant coordinates; but it does sound as though this exercise could in principle be carried out. Note, however, a presupposition we are committed to in so carrying it out. By far the most perspicuous way of accomplishing this is to examine the way quantum theorists actually do carry out calculations of this sort.

One standard procedure is to represent the combined, interacting systems by a Hamiltonian function consisting of two parts—one part representing the undisturbed target atom, the other part representing a potential accounting for the perturbation of this atomic system brought about by its interaction with the incident particle. Time-dependent perturbation theory is then used to find the relative probabilities of various atomic transitions due to inelastic scattering events. The equation representing the interacting systems may or may not have exact solutions, but whether or not it does, in general nonzero probabilities for more than one transition will result. That is, more than one nonzero transition probability will result if the perturbing potential is chosen in anything like a natural way. Bohm's claim is equivalent to the claim that, knowing which transition has occurred, we can always choose a potential of such a form that one transition probability will have value 1.0, the others value zero.

I sketch below a proof of the standard result that a necessary and sufficient condition for the vanishing of all but one transition probability is the vanishing of all but one of a set of corresponding coordinate-integrals involving the perturbing potential, $V = V(q,t)$ for coordinates q and time t. It does seem that with a great deal of ingenuity one could find the appropriate *form* of the potential for each transition. Now we must ask ourselves why the *form* of the potential should change for each transition. Normally, what we expect in repetitions of deterministically describable interaction between particular physical systems is for the form of the potential representing these interactions to be always the same, any difference in the results of the interaction being attributable to differences in the initial values of the coordinates. This normal situation does not suffice for Bohm's purposes. A potential of the form needed to uniquely predict some particular transition will not predict any other. For each "unique" prediction, a potential of a different form is required. To explain the statistical distribution of many inelastic scattering events, we need to assume not merely a distribution of hidden parameters in the sense of positions or momenta, we must also assume that each particular configuration of such variables produces a *qualitatively* different effect.[9]

Two things now become clear. First, we can begin to sympathize with Heisenberg's complaint that Bohm's deterministic theory is really not all that different from the standard indeterministic theory. The complaint, as we have seen, is not accurate; but the force of the complaint is that anyone committed to the presuppositions of Bohm's theory might as well accept indeterminism. If the appearance of determinism can only be saved by a series of retroactive calculations, and calculations based on a physical assumption of incredible strangeness at that, then determinism hardly seems worth the price. Second, the claim that we could choose a perturbing potential of such a form as to rule out all but one transition is now seen to be no objection to an indeterministic construal of quantum theory. We need a potential of a different form for each transition, i.e., we need to make a new and fantastic physical assumption, one not made by the standard quantum theory or, for that matter, any other theory.

There is an exact analogy between the inelastic scattering case just discussed and measurement-interaction cases. Bohm claims to have avoided von Neumann's conclusion because, contrary to von Neumann's premises, Bohm is not limited to a *single* distribution of hidden parameters. Rather, for each particular statistical distribution of measured results, e.g., momentum measurements, a different distribution of hidden parameters is postulated. This move of Bohm's is hardly sufficient. To explain the results of individual measurements—i.e., individual physical interactions—within a statistical distribution of measurements of a given kind, Bohm needs to assume more than a particular distribution of hidden parameters. He needs to assume that each particular configuration representing a member of this distribution produces a qualitatively different effect. Realizing this, we can say that Bohm has succeeded in circumventing von Neumann's conclusion, but only at the price of (implicitly) accepting a presupposition which it never would have occurred to von Neumann to suppose that anyone would accept. Indeed, the "physical" presupposition involved here is more properly regarded as an "unphysical" presupposition.

I now proceed to sketch a proof of the theorem cited above and defend my use of it. Given a two-part Hamiltonian representing the original system plus the perturbation,

$$H\ (q,t) = H_0\ (q) + V\ (q,t),$$

we want to solve the time-dependent Schrödinger equation

$$i\hbar\ \frac{\partial \psi}{\partial t} = H\psi. \tag{1}$$

Suppose the original system is described by

$$H_0\phi_\nu = E_\nu\phi_\nu, \qquad \nu = 0, 1, 2\ldots.$$

Then we can expand $\psi = \psi\ (q,t)$ in terms of the time-dependent eigenfunctions of H_0. That is,

$$\Psi = \Sigma c_\nu\ (t)\ \phi_\nu\ (q) \exp - E_\nu t/\hbar. \tag{2}$$

Substituting equation (2) into equation (1) yields the following set of equations for the coefficients $c_\nu(t)$, where $\mid c_\nu\ (t)\mid^2$ is the probability for transition to ϕ_ν in time t:

$$i\hbar\ \frac{dc_\nu}{dt} = \Sigma_{\nu\neq\mu}\ V\ _{\nu\mu}\ c_\mu\ (t)\ \exp\ (iw_{\nu\mu}t), \qquad \nu = 0, 1, 2\ldots,$$

where $w_{\nu\mu} = \dfrac{E_\nu - E_\mu}{\hbar}$

$$\text{and } V_{\nu\mu} = \int dq\ \phi_\nu\ (q)\ V\ (q,t)\ \phi_\mu\ (q)$$

Boundary condition: $c_\nu(0) = \delta_{\nu_0}$, so

$$c_\nu\ (t) = c_\nu\ (0) + \frac{1}{i\hbar} \int_0^t dt'\ \Sigma_{\mu\neq\nu}\ V\ _{\nu\mu}\ c_\mu\ (t')\ \exp\ iw_{\nu\mu}t'$$

$$= \delta_{\nu_0} + \frac{1}{i\hbar} \int_0^t dt'\ \Sigma_{\mu\neq\nu}\ V\ _{\nu\mu}\ c_\mu\ (t')\ \exp\ iw_{\nu\mu}t'$$

$$= \delta_{\nu_0} + \frac{1}{i\hbar} \Sigma_{\mu\neq\nu} \int_0^t dt'\ V\ _{\nu\mu}\ c_\mu\ (t')\ \exp\ iw_{\nu\mu}t'.$$

Consider a transition between initial state "O" and some final state, say, "1." Then for $t \gtreqqless \tau$ we want

$$c_\nu\ (t) = \delta_{\nu_1}$$

so

$$c_\nu\ (\tau) = \delta_{\nu_1} = \delta_{\nu_0} + \frac{1}{i\hbar} \Sigma_{\mu\neq\nu} \int_0^\tau dt\ v_{\nu\mu}\ c_\mu\ (t)\ \exp\ (iw_{\nu\mu}t)$$

where τ is the time during which the two systems are interacting. During the open interval $0 < t < \tau$ we cannot make any assumptions about the $c_\mu(t)$ values since the original system will be in a mixed or indefinite state, not an eigenstate. That is, in general, all the $c_\mu(t)$ values would be nonzero in this interval.

Since the above set of equations must hold in general, i.e., for arbitrary spectra E_ν and corresponding eigenfunctions ϕ_ν, each integral of the sum of integrals for $\nu \geqq 2$ must be individually zero. Furthermore, since $V_{\nu\mu}$ and $V_{\nu\mu}$ are equal in magnitude, differing only by a phase factor, we are left with

$$\int dt\, V_{10}\, c_0\,(t)\, \exp\,(iw_{10}t) = 1.$$

That is, all the $V_{\nu\mu}$ except V_{10} (and V_{01}) must be of zero magnitude. (To see the above steps it is helpful to write out the simultaneous equations in rectangular array.) This establishes the cited necessary condition. The sufficiency condition is obvious by inspection.

To put the result we have established in proper perspective, a few clarifying remarks are in order. The argument I have given is a reductio ad absurdum, a familiar philosophical tactic. What I have shown is that if we accept Bohm's claim, or something clearly equivalent to it, we must also accept a ridiculous presupposition about physical processes. Some critics might regard this exercise as superfluous, on the grounds that the premise of the reductio argument is patently false to anyone knowledgeable in quantum theory; hence the reductio proof is wasted motion since its conclusion could not be more absurd than its premise. It is pleasant enough to take such criticism in stride, for in effect we are being told we are right and need not produce an argument to support our claim. Still, it seems that the reductio argument is of some potential value since there might be individuals, anxious to accept reinstatement of determinism, who do not find Bohm's claim prima facie absurd but who would regard it with grave suspicion once they realize it commits them to the cited presupposition.

Let us now consider objections of a completely different sort. It might be objected that it is unfair, in a subtle way, to use any of the apparatus of the standard theory in mounting an argument against Bohm's quite different theory. Of course, it is perfectly legitimate to use the *equations* of the standard theory since Bohm takes these over unchanged; so the objection must refer to the methods used in manipulating these equations—e.g., perturbation theory. It might be argued, e.g., that Bohm could in principle calculate the spatio-temporal details of the transition in question directly and so short-circuit the perturbation analysis. What, if anything, follows from this sound claim? No matter how hidden variable theorists actually perform the calculation they will have to choose a potential $V(q,t)$ which ensures that one particular final eigenstate results. Since the equations of the hidden variable theory are those of the standard theory the hidden variable analysis is equivalent to the perturbation analysis I have employed—one analysis will yield the results required by hypothesis if and only if the other will. Once this point is perceived, the charge of unfairness can be laid aside with a clear conscience. It would be equally misguided to object to the expansion of the new, perturbed wave function in terms of the eigenstates of the old, unperturbed Hamiltonian. Once again, even if the new Hamiltonian wave equation has a known exact solution, it will have a unique eigenstate solution if and only if the subsystem described by the old Hamiltonian is constrained, via $V(q,t)$, to making one particular transition.

One final objection: a hidden variable theorist might claim to be able to take the cited presupposition in stride, might claim that, far from being absurd, it is eminently reasonable. After all, the argument goes, why shouldn't we expect a different form for (the coordinate dependence of) the perturbing potential? This is equivalent to saying that the form of the wave function is different for each distribution of coordinates. If the wave function represents an actual physical field, analogous to the electromagnetic field, then why shouldn't the field function change form as the distribution of coordinates changes? Shouldn't we expect that the ψ-field, just as the electromagnetic field, would have to be specified independently, over and above the specification of the configuration of particles? Well, no, we shouldn't expect this. The electromagnetic field must, in general, be specified independently because it is a radiation field, a field capable of energy and momentum absorption, emission, storage, and transmission; the specification of (charged) particle configurations does not serve to specify the electromagnetic field because these configurations do not tell us the net energy and momentum transfer which has taken place previously between the field and the system of particles. Bohm's ψ-field is not advertised as such a radiation field, and in fact Bohm's interpretation presupposes that the ψ-field is *not* a radiation field but rather a field of constant integrated intensity, varying only in spatial form. Accordingly, we can legitimately demand that the ψ-field behave like other nonradiation fields. When we find that the field or, equivalently, the potential is not specified by the configuration of particles, we can justly claim to have isolated an absurd consequence of Bohm's interpretation.

The following sketch of the history and present status of the "hidden variable issue" will indicate why all our conclusions apply quite generally and not merely to Bohm's original (1952) theories. A good clue as to the extremely formidable and abstruse character of this issue is the twenty-year interval between the original (German) publication of von Neumann's theorem and the publication of Bohm's first (two) theories; and the further twelve-to-fourteen-year interval between Bohm's suggestions and clear analyses of just what is allowed and what is proscribed by von Neumann's theorem. Inspired partly by Bohm's apparent "success" and partly by independent considerations, a host of hidden variable theories appeared after 1952. De Broglie (1953) attempted first a theory with "pilot waves" playing a role corresponding to that of Bohm's quantum mechanical potential; and later (1960) a nonlinear quantum theory. Weizel (1953) hypothesized a new particle, the "zeron," to account for the alleged hidden variable effects. Janossy (1953) proposed an altered set of quantum mechanical equations and hence an altered set of solutions, involving damping terms, in an attempt to obtain a quasi-deterministic explanation of measurement interactions. Bohm himself proposed several later hidden variable theories, including one developed in collaboration with Bub (1962).

All the above hidden variable theories, as well as the many others which could be cited, are subject to the same criticism as Bohm's original theories.[10] There is on present evidence no reason whatever to believe any of them true, since each postulates the existence of items we have no reason to suspect exist—pilot waves, new particles, new fields, new interaction forces, etc. This judgment applies to those theories which use the same set of basic equations as the standard quantum theory and, a fortiori, to those with altered equations. What is not obvious is whether the *former* theories are ruled out in advance as intrinsically defective by von Neumann's theorem. It is important to note a possible ambiguity in the phrase 'hidden variables.' Strictly speaking, hidden variables are those whose values at a given time *determine*, in one-one correspondence fashion, the outcome of physical interactions, including measurement interactions. It is not always clear to the reader—and hence probably not to the writers themselves—whether such authors as Blokhintsev (1953; 1968), Alexandrow (1952), Bopp (1947), and Feynes (1952) are defending hidden variable theories in this strict sense or whether they are defending the weaker claim I advocate: that physical systems possess intrinsic properties such as (simultaneous) position and momentum but that the values of these variables do *not* determine interaction outcomes, the inherently indeterministic quantum description corresponding to the actual nature of physical processes. Much confusion has been engendered by the failure of both defenders and opponents of the Copenhagen interpretation to distinguish these two positions. One need not be a hidden variable theorist to defend the weaker claim. Conversely, establishment of this weaker claim provides no support for hidden variable theories.[11]

The restrictions imposed by von Neumann's theorem can best be understood by reference to a number of elegant studies which have appeared since 1957. A crucial assumption in the proof is that the sum of the expectation values for several operators is equal to the expectation value for the operator sum of those several operators. This presupposition is reasonable for *ensembles*—many physical systems measured simultaneously, or the same system measured many different times—but is simply irrelevant for individual measurements on individual systems. A hidden variable theorist might seize on this point and claim that while the cited presupposition is incompatible with the existence of dispersionless states, it is the presupposition, not the existence claim, which should be abandoned. Equivalently, it might be said that only an inapplicable, irrelevant restriction stands in the way of a successful hidden variable theory. The work of Bell (1964; 1966) in part suggests this, but other aspects of his work and his analyses of the work of Gleason (1957) and Jauch and Piron (1963) indicate that all hidden variable theorists are committed to the same sort of absurd consequence which I extracted from Bohm's 1952 theory.[12]

In particular, the Gleason-Bell analyses show that in order to reproduce exactly all the predictions of the standard quantum theory, a hidden variable theory must presuppose that measurements on one of two *causally* separated systems—widely separate in space and not connected by physically measurable fields—instantaneously affect the properties of the other system. Equivalently, we can say that the appearance of determinism can be preserved only by assuming instantaneous action at a distance, by assuming effects for which no physical causes are possible. It is ironic that hidden variable theorists should begin by claiming that identical causes produce identical effects and end by saying that some effects are, in effect, causeless. This position is not, strictly speaking, self-contradictory since there could conceivably be "causes" which do not operate within space and time in the usual way. But such causes are not "physical" causes;[13] they are, quite literally, dei ex machina.

5
Quantum Mechanics and Logic

In this chapter we will examine a second alternative to the Copenhagen interpretation, one based on alleged causal anomalies and other apparent contradictions frequently cited in discussions of quantum mechanical phenomena. The basic argument is that by far the best way of resolving these apparent contradictions is to admit that classical logic does not apply to these phenomena and to "read off" the allegedly correct logic from (the mathematical apparatus of) the quantum mechanical theory that correctly predicts the phenomena in question.[1] Waiving the obvious objections to this program which could be made on general philosophical grounds—objections well made by Fine (1968) and Heelan (1970)—consider the specific claims involved.

It is often said that statement-forms which are equivalent in classical logic—i.e., have the same truth value, either both true or both false—can be shown to be nonequivalent when their interpretations (instances) involve quantum mechanical subject matter. Let S_1 be the statement expressed by "particle system s has momentum p and [either has position q_1 or has position q_2. . .or has position q_n]." Let S_2 be the statement expressed by "particle system s either [has momentum p and position q_1] or [has momentum p and position q_2]. . .or [has momentum p and position q_n]." The two statements are of the form

$$S_1: \quad p \, [q_1 \vee q_2 \vee \ldots q_n]$$

$$S_2: \quad [p \, \& \, q_1] \vee [p \, \& \, q_2] \vee \ldots [p \, \& \, q_n]$$

with '&' for 'and' and 'v' for 'or.' The standard argument is, then, that each of the disjuncts of S_2 is false, since particles do not possess simultaneous position and momentum; hence S_2 is (always) false. But S_1 is equivalent to p, the conjoined disjunction $[q_1$ v q_2 . . . $q_n]$ being always true since by hypothesis there are exactly n possible position values. (The argument does not depend on there being only discretely many values; it is easily generalized to cover the continuous-valued case.) So if p is true, S_1 is true, but in that case S_1 and S_2 have been shown to differ in truth value and so cannot be equivalent.

The above argument rests on both equivocation and misinterpretation of the uncertainty relations. To say that the typical disjunct (p & q_i) is always false, cannot mean that s has no simultaneous position and momentum value. We can always measure, retroactively, the simultaneous momentum and position of a particle by passing it through a narrow aperture and inferring from the position at which it strikes a detector what momentum it must have had when passing through the aperture in order to arrive at the point at which it struck the detector. There is no theoretical limit to the accuracy with which such *past* position-momentum values can be measured.[2] However, such measurements are entirely after the fact; we could not infer the future behavior of the particle from this measurement—that is, the behavior after striking the detector— since the position "measurement" implicit in its striking the detector disturbs the momentum.

Furthermore, if we pass the same particle through the same aperture many times, we will be unable to predict each of the points at which the particle strikes the detector. The location at which such strikes take place, hence the corresponding momentum values at the aperture, will be statistically distributed. A quantum mechanical description enables us to predict this distribution but not the individual trajectories. Analogous remarks apply to the case where s is taken to be a beam of similar particles. The uncertainty relation refers here to the reciprocal uncertainties involved in collimating the beam; the wave function representing the beam has built-in limits regarding such collimation. Attempts to insure that all the particles of the beam have the same position (same position in the xy plane if the beam is travelling in the z direction) such as passing it through a small aperture, will result in the particles of the beam that are passed through the aperture having considerable spread among their momentum values. The smaller the aperture, the larger the spread; the larger the aperture and hence the sloppier the position collimation, the smaller the spread of momenta.

It would be illegitimate to suppose that the above-cited inference presupposes the truth of the classical laws of motion. All that is presupposed is that momentum is conserved in individual processes, a presupposition perfectly consistent with quantum theory and, further, supported by fairly direct experimental evidence—e.g., Compton scattering. Quantum theory neither requires nor forbids momentum (or energy) conservation for individual processes; it requires conservation of the corresponding operator expectation values, but this is a statistical, not an individual, requirement. Nevertheless, because of the cited experimental evidence, quantum theorists do accept the conservation laws for individual processes and even use them as criteria in deciding if certain processes—transitions, scattering events, nuclear reactions, etc.—are possible. One particularly callow error we should avoid is the fallacy of assuming that because the conservation laws, or at least the special cases involving material particle processes, are entailed by the laws of motion, the falsity of the latter entails the falsity of the former.

There is an interesting historical sidelight on this issue. A few years before Heisenberg, Schrödinger, et al. discovered what we now call quantum mechanics, Bohr, Kramers, and Slater (1924) developed a theory about various atomic processes which denied those conservation laws for individual processes, while retaining the statistical conservation laws. Although some of the ideas of their paper proved of great heuristic value in later theoretical developments, their hypothesis of the falsity of the conservation laws was universally rejected, even by Bohr, Kramers, and Slater themselves, largely because of an analysis of Compton scattering experiments. Since then, this rejection has been further supported by a wide variety of other experiments.

Returning to the argument, we see that S_1 refers to a single system s, e.g., a single particle, while S_2 refers, elliptically, to a collection of similar systems. S_1 asserts that s has a definite momentum value and that it simultaneously has some position or other although this position is completely unspecified. In the sense in which each of its disjuncts is false, S_2 asserts that, given a definite momentum value, the dispersion in successive position measurements on s or, equivalently, in simultaneous position measurements on similar systems, will never be zero no matter which position value the spectrum of measured results centers on—q_1 or q_2 . . . or q_n. No position-momentum state of s represents an exactly defined, *reproducible* state. Hence S_1 and S_2 only appear formally equivalent because of ambiguities in their occurring terms; actually they assert two quite distinct statements. Accordingly, any revision of our basic logical principles is completely gratuitous. In the sense in which each disjunct of S_2 is false, S_2 refers to ensembles of particles —many similar particles at the same time or the same particle successively observed many times in indistinguishable circumstances. If we interpret S_2 to refer to an individual path of an individual particle, then the thought experiment described above shows we have good reason to believe that one of the disjuncts of S_2 is true. In no case do we get any failure of the classical equivalence in question. In the sense in which S_1 and S_2 both refer to individual particle paths, both are true. In the sense in which S_1 is true and S_2 false, they represent quite different statements. The point here is that the uncertainty relations, just like the other results entailed by quantum theory, are statistical statements. A little reflection reveals that the theory is agnostic about individual events. Defenders of the Copenhagen interpretation would take this as equivalent to an admission that any claims one way or the other about individual processes are *meaningless*. For obvious reasons, discussed in chapter 2, this is an incredible non sequitur.

The superfluous character of revisions of logic is even more evident upon analysis of arguments based on the diffraction of beams of material particles by crystal lattices, resulting in characteristically wavelike interference lines. Usually, the idealized two-slit thought experiment, rather than actual crystal diffraction experiments, is discussed. One common procedure is to present a classical derivation of the probability distribution of particle strikes upon a screen after the particles have passed through a quasi-periodic (two-slit) lattice. It is then argued that the weakest, least plausible assumption in the derivation—which results in a predicted distribution disagreeing radically with the observed distribution—is the assumption of equivalence between two compound statements of form, respectively,

$$q \text{ and } (r \text{ or } s) \quad \text{and} \quad (q \text{ and } r) \text{ or } (q \text{ and } s).$$

The classical derivation proceeds as follows. Let r and s, respectively, stand for the propositions that a given material particle passes through slits no. 1 and no. 2; let q stand for the proposition that the particle strikes spot x on the screen; let $P(q\&r)$ and $P(q\&s)$, respectively, be the probabilities that the particle strikes the given spot x on the screen assuming it passes through slit no. 1 when no. 2 is closed and vice versa. We can arrange our imagined apparatus so that the probabilities of striking each slit are equal. Since we are only counting particles that reach the screen through one slit or other, the disjunction r or s is always true for particles hitting the screen. Hence

$$P \;(\;(r \text{ or } s) \text{ and } q) = \frac{P \;(\;(r \text{ or } s) \text{ and } q)}{P \;(r \text{ or } s)}$$

$$= \frac{P \;(\;(q \text{ and } r) \text{ or } (q \text{ and } s)\;)}{P \;(r \text{ or } s)}$$

$$= \frac{P \;(q \text{ and } r)}{P \;(r \text{ or } s)} + \frac{P \;(q \text{ and } s)}{P \;(r \text{ or } s)} \; .$$

Since $P \;(r) = P \;(s)$,

$$P \;(r \text{ or } s) = 2 \, P \;(r) = 2 \, P \;(s),$$

so

$$P \;(\;(r \text{ or } s) \text{ and } q) = \tfrac{1}{2} \, P \;(q \text{ and } r) + \tfrac{1}{2} \, P \;(q \text{ and } s),$$

a result which disagrees radically with observation. It is claimed that the "crucial" and "fallacious" step in the above derivation is the expansion of q and $(r$ or $s)$ into $(q$ and $r)$ or $(q$ and $s)$. The only alternatives to rejecting this classical expansion are alleged to be:

(A_1) the assumption that a given particle somehow goes through both slits at once; or

(A_2) the assumption that a given particle "prefers" one slit to the other, but only when no detectors are placed in or near the slits to check this preference; or

(A_3) the assumption that a given particle going through, say, slit no. 1 "knows" that slit no. 2 is open and behaves differently than it would if slit no. 2 were closed.

If choices (A_1)–(A_3), interpreted literally, were the only alternatives, we might, in sheer desperation, entertain the possibility of rejecting the classical logical equivalence in question. Fortunately, choices (A_1)–(A_3) as formulated are hardly exhaustive, though there is a sense in which (A_3), purged of its anthropomorphic aspects, is the correct choice after all. Let us focus on the ambiguous claim that the derived result disagrees radically with observation. In the sense in which the claim is true, it means that if $P(q$ and $r)$ and $P(q$ and $s)$ are each assigned the values they would have if slits 1 and 2, respectively, were open *alone*, then the derived result is incorrect for the case when *both* slits are open simultaneously. Once we realize this, we see that there may be nothing at all wrong with the derivation; if proper values are assigned to $P(q$ and $r)$ and $P(q$ and $s)$, the derived result may be correct. Conversely, in the sense in which the derived result disagrees with observation, this disagreement may not be due to any invalid inference leading to the result but rather to the very dubious presumption that the multiple slit arrangement passes a beam of particles which is qualitatively similar to that passed by each slit acting alone. This dubious presumption is demonstrably false.

Recall that we cannot represent the incident beam in question as a set of particles each having a well-defined, reproducible trajectory. We must represent the incident beam by a wave function with conjugately related position and momentum uncertainties. Given that this representation of the incident beam is appropriate, as opposed to a representation in terms of a set of classically determined individual trajectories, the "interference" phenomena exhibited in the multiple slit thought experiments or the actual crystal lattice diffraction experiments becomes readily explicable.

Landé, a voluminous defender (1955, 1960, 1965), of a particle interpretation of quantum mechanics, long ago realized that the same well-known analysis which provides a particle explanation of crystal lattice diffraction phenomena (Davisson and Germer, Thomson, etc.) also explains slit diffraction, even for the case of only two slits. Landé frequently refers to the pioneering work of Duane (1923) and of Ehrenfest and Epstein (1924, 1927). I propose to establish three points, the first two in this chapter and the third in the next. First, in spite of the important differences between slits and crystals, the same characteristic selective scattering or selective momentum transfer occurs for slits as well as crystals, and even occurs for only two slits. Second, although the key physical insight in my analysis is the same as that of Ehrenfest and Epstein, it is unsatisfactory, for reasons given below, simply to apply their work to explain slit material particle diffraction problems. Third, the same analysis which obviates a "realist" wave interpretation of material particle diffraction also obviates, for exactly the same reasons, the anomalous assumption of probability interference. On this third point I part company with Landé.

Let us dispose of the second point. In their analysis, Ehrenfest and Epstein *assumed* one form of the Bohr-Sommerfeld quantum conditions, where the path integral of the momentum is quantized. Thus their result amounts to converting one quantum condition to another; equivalently, we could say they allowed themselves a large head start on their conclusion. Second, they really did not use what we now call quantum theory at all since they wrote three years before the discoveries of Schrödinger and Heisenberg. Third, they discussed photons, not material particles, and did so in terms of an eclectic mixture of classical wave theory and the above-cited quantum condition. Contrary to popular myth, photons are not particles at all. A quite different explanation of photon interactions is required (see chapter 7). For all these reasons a new analysis, such as the one presented below, is desirable, an analysis which derives a quantum condition for material particles without explicitly assuming any other such condition but rather assuming only the general principles of the Schrödinger theory.

It is instructive to sort out the differences and similarities between slits and crystals in a brief preliminary survey. Any satisfactory interpretation of quantum mechanics must provide an explanation of both crystal and slit phenomena. The crystal experiments provided the first direct confirmation of the de Broglie "matter-wave" hypothesis and, slightly later, of Schrödinger's wave mechanics. Crystal diffraction experiments are literally done every day (somewhere or other), and any generally satisfactory interpretation of quantum mechanics must be able to account for crystal effects. In particular, a particle interpretation must explain how the wavelike interference patterns come about in beams reflected from or transmitted through crystals; this must be done without resort to actual physical waves, spatiotemporal oscillations of measureable physical quantities. Slit phenomena must also be explained by any satisfactory interpretation; even though the few-slit experiments are unperformable idealizations, the results are, all the same, theoretically achievable results which cannot be ignored.

There are important differences between the slit experiments and the crystal experiments. First, the incident beam travels (roughly) *perpendicular* to the slits while in the crystal experiments it is the components of the beam traveling *parallel* to the directions of lattice periodicity that produce the effect in question. Furthermore, the crystal effect is based on the existence of a very large number— virtually an infinite number—of atomic scattering centers along the path of the beam, while the slit effect we are seeking to explain occurs for as few as two slits. In spite of these important differences, there is an important analogy between slits and crystals. Both slit diffraction and crystal diffraction amount to perturbation of an incident beam by a periodic or at least quasi-periodic potential. Both crystal diffraction experiments and slit diffraction experiments appear to exhibit wavelike interference patterns, "probability interference," and the sort of alleged "causal" anomaly which has inspired such desperate proposals as the revisions of logic under discussion and the hidden variable theories discussed in the previous chapter. But the same perturbation-theory analysis, using a periodic perturbing potential or, more precisely, a Fourier-analyzed potential will explain away all these anomalous results for both slits and crystals.

It is a well-known result (see Slater 1968) of the quantum theory of solid-state physics that beams of particles moving along a spatially periodic potential such as that constituted by a crystal lattice will be selectively scattered. Scattering will take place only through certain selected angles; equivalently, momentum transfer will not be continuous but discrete—the particles can gain or lose momentum only in certain discrete amounts when interacting with the lattice. To recall how these results come about, represent the incident beam of particles by a plane wave function

$$\psi_0 = \exp ik^0 x$$

for momentum wave number k^0 and periodicity in the x direction. The effect on the beam of a potential of periodicity a

$$V(x) = V(x+a)$$

is given by the perturbation theory expression,

$$V_{01} = \int_{-\infty}^{\infty} dx \exp - ik^0 x \, V(x) \exp ik^1 x,$$

for the probability (amplitude) of transition from momentum state k^0 to momentum state k^1. If we expand our perfectly periodic potential in a Fourier series

$$V(x) = \Sigma_{-\infty}^{\infty} V_1 \exp i2\pi lx/a, \qquad 1 = 0, \pm 1, \pm 2, \ldots,$$

then

$$V_{01} = \Sigma_1 V_1 \int_{-\infty}^{\infty} dx \exp [ix(k^1 - k^0 + 2\pi l/a)].$$

The integral in the expression for V_{01} is identically zero *except* when the exponent of the exponential function is zero. That is, we only get transitions from the initial states to other states when

$$ix (k^1 - k^0 + 2\pi l/a) = 0$$

or

$$k^1 - k^0 = \Delta k = -2\pi l/a, \qquad 1 = 0, \pm 1, \pm 2, \ldots.$$

So the momentum gained or lost by the beam particles in inter-actions with the lattice does not range over continuous values but rather is quantized. Equivalently, scattering does not take place through all angles but only through selected, discrete angular values. Clearly, when a beam reflected from such a periodic lattice strikes a detecting "screen," the resulting pattern will exhibit a discrete-line pattern, an "interference" pattern. Similarly, if a beam is passed through a "thin" crystal—the thinnest crystals are fifty atomic layers thick, which is virtually an infinite lattice—the result-ing pattern will again be a line pattern, an "interference" pattern.

In view of these well-known results we can see that there is nothing anomalous in the effect which crystal lattices (or multiple slits) have on incident beams. Once the incident beam is properly represented, the effect is exactly the one to be expected. Thus, in tracking down the source of puzzlement, surprise, anomaly, etc. in such phenomena we should realize that it lies entirely in the de-scription of the incident beam. There is nothing surprising in the fact that a periodic arrangement of material produces a periodic effect on such an incident beam. Indeed, with benefit of hindsight, we can see that the opposite assumption, that a periodic arrange-ment will behave in completely unperiodic fashion, is most implausible.

We now turn from crystals to slits. Consider a beam of particles incident on a multiple slit arrangement periodic in the y direction with slit spacing a and slit width δ. Ignore the z direction and treat the problem as two-dimensional. In general, the incident beam is represented by

$$\psi(x,y) = \exp i\mathbf{k}\cdot\mathbf{r} = \exp i(k_x x + k_y y).$$

That is, the beam is represented by a plane wave with nonzero x and y components. Each individual particle of the beam would be represented by a wave packet or superposition of plane waves; but if the (time-integrated) number of particles in the beam is large enough, the total set of wave packets can be combined to yield a single plane wave. This assumption is exactly the same as is made in the one-dimensional crystal diffraction analysis where the beam is traveling parallel to the direction of periodicity.

The effect on the incident beam of the periodic slit arrangement is the effect of a periodic perturbing potential $V(y) = V(y + a)$. The probability of transition between initial state zero and final state 1 is given by the low-order perturbation theory

$$V_{01} = \int_{-\infty}^{\infty} dy \; \psi^*_1(x,y) \, V_y \psi_0(x,y).$$

This low-order approximation is accurate because the potential represented by the slit arrangement, though large, acts only for a short time—only at the instant the beam passes the slits. The time-integrated effect is small.

We know we can always expand the periodic potential in a Fourier series

$$V(y) = V(y + a) = \Sigma_l V_l \exp i2\pi ly/a, \qquad l = 0, \pm1, \pm2, \ldots,$$

then

$$V_{01} = \Sigma_l V_l \int_{-\infty}^{\infty} dy \exp [-i \,(k_x^1 x + k_y^1 y)] \exp iy2\pi l/a \exp$$
$$[i(k_x^0 x + k_y^0 y)]$$
$$= \Sigma_l V_l \exp [ix(k_x^0 - k_x^1)] \int_{-\infty}^{\infty} dy \exp [iy \,(k_y^0 - k_y^1 + 2\pi l/a)].$$

The y integral is zero except for the contributions which occur when the integral is a constant; i.e., when

$$\exp [iy \,(k_y^0 - k_y^1 + 2\pi l/a)] = \text{const.} = e^0 = 1$$

or

$$k_y^0 - k_y^1 + 2\pi l/a = 0.$$

So we see that the y component of momentum does not change in continuous fashion, but rather the changes are quantized according to

$$\Delta p_y = \hbar\Delta k_y = \hbar 2\pi l/a = hl/a, \qquad l = 0, \pm 1, \pm 2, \ldots .$$

In the above calculation the matrix element V_{01}, related to the transition probability P_{01} by const. $P_{01} = |V_{01}|^2$, is a small quantity since the coordinate integral involved only receives contributions from selected pairs of initial and final momenta or, equivalently, from a deltalike function at $y = 0$. Accordingly, if the perturbed wave function is to differ noticeably from the incident beam function, if a noticeable "interference" pattern is to be seen over the plane-wave background pattern, then since the change in the wave function is proportional to

$$\frac{V_{01}}{E_0 - E_1}$$

the condition $E_0 \cong E_1$, i.e., the elastic scattering condition, must be satisfied approximately, though not exactly. This means that the mass of the slit arrangement must be very large compared to the mass of the beam particles, and the slits must be rigidly coupled so as to act as a unit. (Rigidity is required in any case for the slit arrangement to constitute a truly periodic potential.)

The elastic scattering condition is equivalent to the condition

$$|\mathbf{k}_0|^2 \cong |\mathbf{k}_1|^2$$

or

$$\begin{aligned}
(k_x{}^0)^2 + (k_y{}^0)^2 &\cong (k_x{}^1)^2 + (k_y{}^1)^2 \\
&= (k_x{}^1)^2 + (k_y{}^0 + 2\pi l/a)^2 \\
&= (k_x{}^1)^2 + (k_y{}^0)^2 + 2k_y{}^0 2\pi l/a + (2\pi l/a)^2.
\end{aligned}$$

Assuming the x component is unaffected by the periodic y potential, we have

$$\mathbf{k}^0 \cdot \hat{\mathbf{j}} \cong 1/2(2\pi l/a).$$

That is, the projection or y component of the initial momentum must be one-half the magnitude of one of the constants characterizing a particular term in the Fourier expansion of the potential. This is a special case of a familiar result from crystal lattice diffraction analysis.

Now let us see what our results mean in terms of the pattern formed on a detecting device by the transmitted beam. If the incident beam is "monochromatic," i.e., with a well-defined momentum (**k**) in both magnitude and direction, then scattering or diffracted beams, whether transmitted or reflected, will be observed only in specific angular directions, only in directions corresponding to scattered vectors

$$\mathbf{k}^1 = \hat{\imath}\, k_x{}^1 + \hat{\jmath}\, k_y{}^1$$
$$= \hat{\imath}\, k_x{}^0 + \hat{\jmath}\, (k_y{}^0 + 2\pi l/a), \qquad 1 = 0, 1, 2, \ldots.$$

The detector pattern formed will not be that formed by the superposition of beams with continuous ranges of y component momentum values passed by each slit. Rather, each slit passes particles with particular momentum components and reflects others. Each particle passes through one particular slit, but the slit action is not that of an isolated slit; the multiple slit arrangement acts as a unit.

There are some obvious objections to the foregoing analysis. First, we have explicitly assumed the incident beam to have a y component of momentum. But the multiple slit phenomenon we are attempting to explain, without recourse to the standard interpretation, occurs even for beams of the form

$$\psi = \exp ik_x x$$

beams with well-defined nonzero x component and a well-defined zero y component of momentum. Our explanation depends on the existence of y components of momentum. How can we invoke this explanation for beams traveling exactly in the x direction? It is the beam, i.e., a statistical assemblage, and not the individual particles, which is well-defined in momentum. When we showed above that scattering, whether forward scattering into the transmitted beam or backscattering into the reflected beam, occurs in selected angular directions, this result was a statistical result. The proper wavefunctional representation for an individual particle is

$$\psi(x,y) = \Sigma_{\mathbf{k}'}\, A(\mathbf{k}')\, \exp\, [i\mathbf{k}' \cdot (\hat{\imath}x + \hat{\jmath}y)].$$

Each particle is represented as a superposition of plane waves, in order to localize the particle to within some given (x,y) region. So long as the particle is at all localized, it will appear to have some dispersion in its momentum components. In particular, so long as the y coordinate is to the slightest extent restricted there will be some "dispersion" in the y component of momentum. Conversely, if the y component is fixed exactly, e.g., $k_y = 0$, then the y coordinate is completely unknown and there is no more probability of the particle striking a finite slit arrangement than there is of its passing on the other side of the moon. If we restrict the y coordinate of the beam, then we will have introduced significant "dispersion" in the particles' y components of momentum. This is equivalent to saying that in a beam containing a statistically significant number of particles, some of these particles will have nonzero y components of momentum.

Hence, our answer to the first objection is that any particle with a probability worth mentioning of striking the slit arrangement will likely have a y component of momentum. Any beam which can produce the phenomenon we are concerned about, multiple slit diffraction, will contain large numbers of particles with y components of momentum. This answer can now be restated specifically in terms of the beam itself. Suppose we want to represent a beam traveling in the x direction, one with $k_y = 0$ and $k_x = k_x$, where these are exact values, with allegedly no dispersion. The beam is represented by

$$\psi_{\text{beam}}(x,y) = \exp(ik_x x),$$

i.e., as a plane wave. How is this plane wave description justified? The beam function is arrived at by superposing the wave packets for each particle of the beam, each packet in turn being a superposition of plane waves. Thus, for particle index m,

$$\psi_{\text{beam}}(x,y) = \Sigma_m \Sigma_{\mathbf{k}'} A_m(\mathbf{k}') \exp[i\mathbf{k} \cdot (\hat{\imath}x + \hat{\jmath}y)]$$
$$= \exp(ik_x x).$$

The last equality holds only if we can find sets of $A_m(\mathbf{k}')$ such that the y dependence of the beam wave function is suppressed. What does this mean physically? Since the "waves" in question are not actual physical waves, i.e., since the wave function does not represent the spatiotemporal oscillations of any physically measurable quantity, we are not actually *canceling* out momenta, as, e.g., by destructive wave interference. Rather, we are *averaging* out the momenta. We can arrange a well-collimated beam with zero average

Quantum Mechanics and Logic

or expectation value for the y component of momentum, but the fact remains that many of the individual particles have y components of momentum. And this is all that is required for our explanation to be invoked. Once again, we should notice that if the plane-wave beam function

$$\psi_{\text{beam}}(x,y) = \exp ik_x x$$

is taken literally, taken to represent a beam with $k_y = 0$, and no dispersion in k_y, then there is no nonnegligible probability of the beam striking the slits.

The second objection we must deal with concerns our tacit assumption of a large number of slits involved in the diffraction phenomena. In obtaining the selective scattering result, we relied on the vanishing of a certain integral, and this assumption amounts to assuming integration over a large dimension compared to a unit dimension associated with a single slit. (Equivalently, it amounts to assuming boundary effects are neglible compared to periodic effects far from the boundaries.) But the theoretical effect we are attempting to explain occurs even for only two slits. Accordingly, we must show that our explanation is suitable even for this special case, which is quasi-periodic at best and not a case involving a strikingly obvious periodicity. Also, it is plausible to require that our explanation show why the selective scattering result does not occur for a single slit though it does occur for any number of slits greater than one. It turns out that these requirements can be met in relatively straightforward fashion if we reflect on the Fourier analysis involved in the problem.

For many slits, the potential closely approximates a perfectly periodic potential, and this justifies our use of the Fourier *series* expansion

$$V(y) = \Sigma_{-\infty}^{\infty} V_1 \exp i2\pi yl/a, \tag{1}$$

where

$$V_1 = 1/a \int_0^a dy \ V(y) \exp -i2\pi yl/a, \tag{2}$$

But for the aperiodic one-slit case and the quasi-periodic two-slit case which do not at all approximate a perfectly periodic potential we must use instead a Fourier *integral* expansion:

$$V(y) = \int_{-\infty}^{\infty} dl \ V_1 \exp i2\pi yl, \tag{3}$$

113

where

$$V_1 = \int_{-\infty}^{\infty} dy \, V(y) \exp - i2\pi yl. \tag{4}$$

Substituting equations (2) and (4) into equations (1) and (3), respectively, and these in turn into the perturbation expression for transition probability amplitude yields

$$V_{01} = \int_{-\infty}^{\infty} dl \int_{-\infty}^{\infty} dy \, \psi_1^*(x,y) \, [\int_{-\infty}^{\infty} dy' V(y') \exp - i2\pi ly']$$

$$\exp i2\pi ly \psi_0(x,y) \tag{5}$$

for the aperiodic or quasi-periodic case and

$$V_{01} = \Sigma_{-\infty}^{\infty} \int_{-\infty}^{\infty} dy \, \psi_1^*(x,y)$$

$$[1/a \int_0^a dy' V(y') \exp -i2\pi ly'/a] \, \exp i2\pi ly/a \, \psi_0(x,y) \tag{6}$$

for the perfectly periodic case. In equations (5) and (6) the wave functions, as before, are

$$\psi_0 = \exp [i(k_x^0 x + k_y^0 y)]$$

and

$$\psi_1^* = \exp [-i(k_x^1 x + k_y^1 y)].$$

Since the Fourier integral expansion of equation (5) is a completely general result, valid for any potential function $V(y)$, no matter what the form, equation (5) should be equivalent to equation (6) for the special case of a perfectly periodic $V(y)$. The best way to proceed in displaying this equivalence is to notice that for the slit problem it is the slits, rather than the opaque material between slits, which are the active agency so far as scattering into the transmitted beam is concerned. Hence, $V(y')$ in the inner y integral of equation (5) takes on nonzero values only in the narrow y regions occupied by the slits. In setting up the slit problem to derive the result we are endeavoring to explain—the characteristic "interference" pattern—the standard assumption is that $a \gg \delta$, the slit spacing is very large compared to the slit width. This is assumed because we are most interested in the low momentum exchanges, i.e., in the forward scattering, which favors large a. Furthermore, this assumption is required to guarantee "coherent" beams from each slit and, more important, to guarantee significant "phase" differences in the beams arriving at distant common points from several slits. But with this assumption about slit widths and spacings, $V(y')$ can be represented as a series of delta functions.

114

That is, the only nonzero contributions to the y' integral come at $y' = \pm na$, $n = 0, 1, 2 \ldots$. Thus the y' integral is really just an infinite series. But replacing the y' integral by a series will not show the connection of equations (5) and (6). So instead of replacing the continuous y' integration by a discrete sum, we replace the l integration, the Fourier term index integration, by a discrete sum and integrate the continuous y' value as before. This trick works because the only l dependence in either inner or outer integral involves the product (ly). So restricting l to values 0, ± 1, $\pm 2, \ldots$ and integrating over y'/a is equivalent to integrating over continuous l and summing over $y' = 0$, $\pm a$, $\pm 2a, \ldots$. Notice that y' is a dummy variable in the inner integral, which is a function only of its limits and its parameter l; hence, when we restrict l to discrete values in the inner integral we must do so in the outer l integral also.

So when we employ the cited trick, we see that equations (5) and (6) are equivalent for the special case of perfectly periodic $V(y')$. But now we notice that the same trick can be employed even if $V(y')$ is not perfectly periodic. So long as we are dealing with narrow slits ($a \gg \delta$), we can replace the integral (5) by a sum of the form (6). The difference between the multiple slit (virtually infinite lattice) case and the few slit cases now shows up as the difference in the number of terms in the l series expansion which we obtain. Since we are trading y' summation, l integration for y' integration, l summation the number of l terms we get is just the number of slits—the number of nonzero $V(y')$ locations. What does this mean in terms of selective scattering or selective momentum transfer? It means that no matter how many slits there are we always have the familiar constraint

$$k_y^0 - k_y^1 + 2\pi l/a = 0,$$

but the range of l is restricted to

$$l = 0, \pm 1, \pm 2, \ldots, \pm(n-1)$$

for n slits. In particular, for two slits we still get selective momentum transfer. In fact the transfers are restricted to only two possible values,

$$\Delta k_y = \pm 1/a.$$

For the single-slit case the restriction is even more severe:

$$\Delta k_y = 0.$$

Thus we see that the single-slit case represents the most selective scattering situation of all, but since there is, on the average, no momentum transfer the distinction between discrete and continuous ranges of transfer breaks down!

We have shown that an explanation of anomalous "interference" patterns for particle beams diffracted through multiple slits can be given in terms of selective scattering or selective momentum transfer and that this explanation works even for two slits. For the single slit we do not get an "interference" pattern, but this is because the momentum transfer is zero and there is then no difference between discrete and continuous transfer. One other point is in order here. The approach to the limit of continuous scattering or continuous momentum transfer for slits is *not* in the direction of fewer and fewer slits. The single-slit case or, equivalently, the case where a is large compared to the beam collimation is the most restrictive of all. Rather, the continuous limit is obtained by taking a to be very small, i.e., a large number of slits *per unit length*. Then $V(y')$ approaches a continuous function, and the scattering is continuous. In this respect the slit diffraction phenomenon is just opposite to the crystal diffraction phenomenon where the continuous scattering limit occurs for large distances between scattering centers, or very few atoms per unit length. Obviously, this difference stems from the fact that the incident beam is, roughly, perpendicular to the slits but parallel to the crystal lattice periodicity direction. In the crystal the distance between atoms must be small for the selective scattering effect to occur; otherwise the beam can recover its asymptotic form in between scattering centers. But the beam strikes all the slits simultaneously; here, we want wide spacing of narrow slits in order to constitute a potential with sharp nonzero values at a few locations and zero values elsewhere. To complete the argument, we should solve explicitly for the transmitted beam wave function

$$\psi_{\text{transmitted}} \cong \psi_{\text{unperturbed}} + \Sigma_0^N \frac{V_{10}}{E_1 - E_0} \psi_1,$$

where

$E_1 - E_0 \cong$ const. because of the elastic scattering condition and

$$\psi_1 = \exp(ixk_x{}^0) \exp[iy(k_y{}^0 + 2\pi l/a)]$$

$$l = 0, \pm 1, \ldots \pm (N-1)$$
$$\text{for } N \text{ slits}$$

Quantum Mechanics and Logic

$$V_{10} \propto V(y) \cong \begin{cases} \delta\,(la) & \text{for odd } N \\ \delta\,[(1 + 1/2)a] & \text{for even } N. \end{cases}$$

More explicitly,

$$\delta\,(\xi) = \exp\,[-(y - \xi/2)^2/2\delta^2]$$

for slit width δ.

For one slit

$$\psi_{\text{transmitted}} \cong \delta(0)\,\exp\,(ixk_x{}^0)\,\exp\,(iyk_y{}^0)$$

$$P(X,y) = \delta^2(0).$$

For two slits

$$\psi_{\text{transmitted}} \cong V_{-10}\,\psi_{-1} + V_{10}\,\psi_1$$

$$= \exp\,(ixk_x)\,\{\delta\,(-a/2)\,\exp\,[iy(k_y{}^0 - 2\pi/a)] +$$

$$\delta\,(a/2)\,\exp\,[iy(k_y{}^0 + 2\pi/a)]\}$$

$$P(X,y) = \delta\,(-a/2) + \delta\,(a/2) +$$

$$\delta\,(-a/2)\,\delta\,(a/2)\,\exp\,[iy(k_y{}^0 + 2\pi/a)]\,\exp\,[-iy(k_y{}^0 - 2\pi/a)] +$$

$$\delta\,(a/2)\,\delta\,(-a/2)\,\exp\,[-iy(k_y{}^0 + 2\pi/a)]\,\exp\,[iy(k_y{}^0 - 2\pi/a)]$$

$$= \delta\,(-a/2) + \delta\,(a/2) + 2\delta\,(a/2)\,\delta\,(-a/2)\,\cos\,4\pi y/a.$$

So for the two-slit case we get a cosine term superimposed on the δ functions; for one-slit a symmetrical δ function. This is exactly the same result obtained in the standard "wave" analysis of the slit problem. (Cf. Tomonaga [1966], 2: 236–38.)

The "ψ unperturbed" term drops out because the transmitted beam consists entirely of particles scattered into this beam by the perturbation, i.e., by the slits. It is imperative that we realize our assumption of infinitesimal slits is a mere convenience. If the slit width is assumed finite (of course it must still be small compared to slit spacing, else the effect we are being asked to explain will not occur), then a large fraction of the particles of the beam will stream through unperturbed. The resulting detector pattern will consist of an interference pattern, produced by those particles which collide with the slit edges, superimposed on the unperturbed, plane wave "background," produced by the particles which stream through. Thus we would end up with a nonzero "ψ unperturbed" term in the function for the transmitted beam. Although the calculation of the relative weighting of the perturbed and unperturbed terms might be a nasty exercise, best left to the diligent reader, no new physics would be introduced. Notice that the consideration we have been discussing only arises in the idealized "thought experiment" with slits. Since it is a thought experiment, we can imagine that only impact forces are allowed, thus the only interactions occur when particles strike slit edges. For the *actual* crystal experiments, of course, there are always well-known fields acting on the incident electrons—or protons or neutrons, etc.—so that the distinction between finite and infinitesimal slit width is of no interest.[3]

Let us connect the results of the correct analyses sketched above with the above-cited alternatives (A_1)–(A_3). Clearly, (A_1) is false; it is conceptually impossible given the meaning of 'particle.' (A_2) is frightfully vague but I suppose what is intended is the statement that a given particle changes its direction (and/or its speed) even though no force is acting on it, the change being sufficient to prevent it from reaching one of the "forbidden" regions on the detecting screen. This alternative is false because it involves violations of conservation laws. (A_3) is a conjunction, the first half of which is an anthropomorphism not deserving further mention, but the second half is in some sense correct—the given particle does "behave" differently when passed through a double-slit arrangement than when passed through an arrangement with one slit closed, though putting the matter in terms of the particle's behavior does not give quite the right emphasis. Rather, we should say that the slit arrangement acts as an extended unit and, not unexpectedly, it acts differently when one slit is open compared to the case where both slits are open. The space distribution of the perturbing potential represented by the slit arrangement is quite different in the two cases.

No revision of logic is necessary to resolve two much-discussed quantum mechanical "anomalies." Furthermore, acceptance of a revised logic would hardly be sufficient to resolve these anomalies in anything like a heuristically satisfying way. To point out that erroneous results would not be derivable if we refuse to sanction various classical inferences, is not to show us how to derive the correct results. Before we can isolate the errors in classical derivations we must have the correct explanations available for comparison.

In view of the unattractive character of revisions of logic as solutions to quantum mechanical anomalies, we might wonder why anyone would be tempted to propound them. There is a trade involved in the proposals under discussion—with a revision of logic as the price for resolving the so-called "causal anomalies" of quantum mechanics. Reichenbach, who earlier advocated a more radical revision involving rejection of the law of excluded middle and acceptance of a three-valued logic, was quite explicit on this point. He repeatedly told us that the only justification for a revised logic was that it enabled us to avoid these causal anomalies (Reichenbach 1946, pp. 32–44).

This way of speaking is misleading; given the quantum mechanical explanation of the crystal scattering experiments, as sketched above, the anomaly is more accurately labeled 'deterministic' rather than 'causal.' There is nothing acausal about the "interference" effect the crystal lattice or multiple slit produces by acting on the incident beam. The periodic potential which the lattice represents is capable of producing this effect and does in fact (always) produce it. Hence it would be misleading to refuse to label the process a causal one. The same judgment applies if we consider the particles of the beam incident one at a time upon the lattice. The lattice will transmit particles with certain momentum components and reflect those with other components; again, nothing "acausal" is involved. However, there is something appropriately labeled 'indeterministic' involved, in that the individual paths of these individual particles are not reproducible; hence the individual paths are not predictable but only inferrible after the fact. The very description of the incident beam—i.e., the wave functional description—which makes the correct explanation of the phenomenon possible has this indeterminism built into it. Any attempt to remove this indeterminism will also remove the basis for the correct explanation. Hence any trade which rejects classical logic in order to preserve determinism is a trade in which we lose twice on the same transaction; we lose (part of) classical logic, which is near and dear to our conceptual hearts, and we also lose indeterminism, which is necessary for the correct physical explanation of the phenomenon we were worried about in the first place.

6
A Defense of Particle Interpretations

Let us review some of our results. Following an exposition of the standard Copenhagen interpretation, it was shown that the two central claims of that interpretation are mutually inconsistent. It is inconsistent to claim that classical mechanics is a special, limiting case of quantum mechanics in any informative sense and also to claim that various key terms common to both theories differ in meaning in the two theories. This inconsistency could be avoided if we hold that the necessary and sufficient conditions for correct application of the term 'particle' are the same in both theories. This position is tenable because of the previously unemphasized separability of the central idea of classical mechanics into two ideas. We can unequivocally label items 'particles' provided they *have* simultaneous position and momentum, even though we admit there are no deterministic laws of motion uniquely relating earlier and later position-momentum pairs, no law describing the individual paths of individual particles. On the basis of a very simple and well-known thought experiment, we have good reason to believe that particles have simultaneous position and momentum.

The upshot so far is that if we take indeterminism seriously, if we genuinely renounce determinism, then it is superfluous, redundant, and gratuitous also to renounce the concept of the path of a particle. The force of this point is illustrated in the famous multiple-slit diffraction experiment, an idealized analogue of the actual crystal-lattice diffraction experiments. That discussion served two main purposes. First, it showed that proposed revisions of logic are misguided since the correct explanation of the anomalous "interference" patterns employs classical logic, and the incorrectness of the classical mechanical prediction is not due to its employment of classical logic but rather to the error in a basic *physical* presupposition. Second, the wavelike interference pattern was shown to be explicable as a buildup of individual particle strikes, the pattern naturally resulting from the effect on the incident beam of the periodic potential constituted by the lattice, provided we represent the incident beam by a statistically interpreted wave function. An indeterministic particle interpretation is necessary and sufficient for a correct explanation of this particular anomaly. Conversely, an interpretation in which the wave function is taken to represent the actual spatiotemporal oscillations of some physical quantity is neither necessary nor sufficient.

The conclusions reached in the analysis of the crystal diffraction experiments can be generalized to other quantum mechanical phenomena exhibiting alleged wavelike properties, causal anomalies, or breakdowns in standard rules of logic inference. Scattering cross-section anomalies and tunneling effects are familiar examples. Readers knowledgeable in quantum theory can verify for themselves that for these phenomena, just as for the crystal-lattice diffraction experiments, there are well-known particle explanations of the anomalies. I do not claim to have "proved" that particle explanations are universally satisfactory. I have merely examined one phenomenon frequently cited as a test case and exhibited an analysis whose appropriate analogues can be employed in other cases.

Putting aside until chapter 7 relativistic and field theory considerations, what remains to be shown concerning particle interpretations is that various questions which have frequently been cited as requiring answers from defenders of particle interpretations either: have already been answered, at least implicitly, by previous analyses; or else are seen upon examination to be incoherent questions, questions based ultimately on self-contradictory presuppositions.

Consider three especially prominent questions. First, why do some physical systems have discrete rather than continuous spectra for values of certain variables? Second, why are there reciprocal uncertainties between (the values of) certain variables, e.g., position and momentum, energy and time, etc.? Third, why are disjunctive probabilities obtained by, e.g., squaring the sum of the amplitude of wave functions rather than summing the squares of the amplitudes—i.e., summing the probabilities themselves? These three questions are chosen for discussion largely by elimination, as the major questions not already given explicit answers in earlier chapters.

The question about discrete spectra seems to arise naturally, largely because, as it happens, most nonprofessional authors are familiar with classical wave phenomena exhibiting discrete spectra but few are familiar with classical particle phenomena exhibiting corresponding discrete, say, energy spectra. Because of this relatively limited and unrepresentative background knowledge of classical theory, many who write on the interpretation of quantum mechanics demand to know how particle interpretations can handle the discrete spectra effects while taking it for granted that such effects are easily accommodated within realist wave interpretations.

The question about the uncertainty relations arises largely because very few authors, professional or otherwise, have noticed—let alone fully grasped and accepted—the separation of the central idea of classical mechanics which I emphasized in chapter 2. If it is asserted that material object systems described by quantum mechanics can be represented as particles, as possessing simultaneous position and momentum, then most authors wrongly presume this in itself entails that the individual paths of these particles are in principle determinable, that there is some set of deterministic laws governing these paths. Once this illegitimate entailment is invoked, it becomes difficult to see why there should be any minimal reciprocal uncertainty in the specifiability of simultaneous position and momentum values.

The question concerning seemingly nonstandard probability manipulations is particularly pressing for indeterministic particle theorists compared to, say, hidden variable theorists or realist wave theorists. The latter two groups defend deterministic theories which, if true, could in principle avoid probabilistic analyses of physical phenomena and thereby avoid any accompanying anomalies in the probability manipulations. Indeterministic particle theorists hold that probabilistic analyses are not mere temporary expedients but reflect the basic nature of physical processes; hence it is incumbent upon us to explain away any anomaly arising in these analyses.

The problem of solving the basic equations of quantum mechanics—the Schrödinger wave equations or their mathematical equivalents—is an *eigenvalue* problem. Whenever we are faced with the problem of solving, subject to boundary conditions, a differential equation containing a parameter—say, v—such that physically possible solutions exist only for certain values of v, then we are faced with an eigenvalue problem. The values of v for which physically acceptable solutions exist are called eigenvalues, and the corresponding solutions for the dependent variable of the differential equation as a function of the independent variable(s) are called eigenfunctions. It is a well-known property of such eigenvalue problems that finite or bounded systems—those for which the dependent variable has value zero except for values of the independent variable(s) less than or equal to some limiting value(s)—have discrete sets of eigenvalues and, correspondingly, discrete sets of eigenfunctions. The solution of one form of the Schrödinger equations, subject to some appropriate boundary conditions, represents such a familiar eigenvalue problem with the energy variable playing the role of the eigenvalue parameter, v, and the wave function, ψ, being an eigenfunction of the independent variable(s)—i.e., of the generalized coordinates, q, or the generalized momenta, p. In particular, the wave equation representing atomic systems—electrons bound to atomic nuclei, i.e., electrons existing in stable "orbits" in relatively close proximity to these nuclei—will have discrete eigenfunctions and corresponding discrete energy eigenvalues.

The above may be felt to be an unsatisfactory reply to the first of the cited questions. What has been pointed out is no more than the fact that the mathematical analysis of bound state problems in quantum mechanics is formally or mathematically equivalent to that used in solving various classical problems. What about the physics of the situation? No explanation has been given of why systems of particles should display, e.g., discrete energy spectra; and, in fact, classical eigenvalue problems characteristically arise in the analysis of wave phenomena—vibrating string or drumheads, etc. So, far from explaining away a difficulty for particle interpretations, haven't we merely shown why wave interpretations of atomic phenomena are so natural? Specifically, haven't we shown why an "explanation" of atomic structure in terms of (three-dimensional) standing wave patterns is so plausible, the discrete values of the wavelengths involved being necessary to insure the stability of the patterns? In answer, we need to sort out and settle three distinct issues.

First, consider the claim that discrete eigenvalue problems are characteristically associated with wave phenomena. If "characteristically" here means "most frequently" or "typically" then the claim is perhaps true but far too weak. The stronger claim that, classically, systems of particles are *never* analyzed in such a way that discrete eigenvalue problems arise is simply false. Discrete eigenvalue problems for systems of particles arise in neutron transport theory,[1] in normal mode analysis of (massive) coupled oscillators, and in electrical circuit theory analysis, all branches of classical physics— of classical statistical mechanics, classical mechanics, and the atomistic theory of electricity, respectively.

There is a second misleading claim involved in the above-cited objections: that wave interpretations provide satisfactory, as well as "natural," explanations of atomic energy states and transitions between such states. Consider the popular heuristic "explanation" of the existence of discrete atomic energy levels and corresponding discrete energies of emitted radiation:

Suppose an electron described a closed, stable path within an atom. If we represent this process as a wave propagation in a continuous field of mass (and charge), then in order for the wave process to be in a stable condition the (closed) path length should be equal to an integer multiple of the wavelength, else the wave front cannot be smoothly joined to the tail of the wave upon completion of a given cycle. Hence for generalized (space) coordinate q we have

$$\int_0^{2\pi} dq = n\lambda, \qquad n = 1, 2, 3 \dots$$

The use of the de Broglie-Einstein relation

$$p = h/\lambda$$

immediately suggests the general condition

$$\int_0^{2\pi} p\,dq = nh, \qquad n = 1, 2, 3 \dots,$$

which is one form of Bohr's quantum condition.

We should resist the temptation to conclude from the above that wave interpretations provide a natural explanation of discrete—i.e., quantized—values of various important dynamical variables, notably action, energy, etc. In the above "derivation" of the quantum condition for the action—i.e., the momentum path integral—we assumed that the *particle* property momentum was related to the wave process via one of the pair of fundamental quantum relations

$$h = E/\nu = p\lambda. \qquad (6.1)$$

Without this assumption, the wave theory in question is powerless to explain anything. Similarly, in obtaining the correct expression for atomic energy states, the wave theory proceeds by finding the "proper" or "characteristic" or "fundamental" oscillations of the field of matter and then relating the frequencies (ν) of these oscillations to the corresponding energies (E) via the other of the above pair of fundamental quantum relations. Before the introduction of this quantum relation there is no reason whatever to interpret the product 'h times the eigenvalue parameter ν' as an energy term. Indeed, without the introduction of this relation we could just as well suppose that the energy of the system being described takes on continuous values. (Even more obviously, why should all experiments indicate the *mass* of the system to behave as though composed of discrete units?)

The situation is quite different with regard to the Schrödinger equation or its mathematical equivalents;[2] these equations are particle equations from the start, the eigenvalue parameter E being interpreted as the total energy of the system of particles and the generalized coordinates (q) and momenta (p) referring to properties of these discrete particles. Strictly speaking, the very attempt to associate the solution of the Schrödinger equation with field amplitudes, i.e., with oscillations in continuous media, only makes sense in the special cases where the particles of the system represented are all mutually independent. For then the appropriate "space" in which the temporal development of the wave function unfolds is our actual, physical, three-dimensional space. For systems involving interactions among the constituent particles, the space in which the time development of the wave function is traced is an abstract, multidimensional—$3N$ for N interacting particles—space. Thus realist wave interpretations are limited from the outset to a very special class of cases.

It is the above-cited objections to realist wave interpretations which are the decisive ones. Some popular objections are so weak as to constitute no objections at all. To point out that the wave function takes on complex values is to make no objection at all unless we forget that, classically, electromagnetic waves are represented by a complex function. Indeed, any process involving two distinct oscillations which are in some phase relation to each other is conveniently represented by a complex-valued function.

The third issue involved in the cited objections hangs on the interpretation to be given to the demand for an "explanation" of the existence of such things as discrete energy levels. In one sense, of course, we can give a very straightforward explanation. Such discrete levels are predicted by the Schrödinger equation, an equation which we have good reason to believe is true because it also successfully predicts a wide variety of *other* results. Furthermore, as we have already noted, this equation is of such a form that discrete eigenvalue solutions are to be expected for problems with certain boundary conditions.

What better "explanation" than this could we give? Could we be asking why we should suppose the Schrödinger equation is true? We have already answered this question; we have pointed out that our reason for supposing it is true is that it is supported by wide and varied evidence, including the discrete energy levels it successfully predicts. Are we then asking why the evidence is true? If we persist in this course of questioning we will soon reach the stage where we ask, "Why are things the way they are rather than some other way?" Such questions are unanswerable. Explanation must begin, and end, somewhere. If we take some things as true we can, if we are very fortunate, explain everything else. But we could not possibly explain everything at once, for to do so would presuppose both that something be taken as true for explanatory purposes and also that nothing be taken as true until explained. Questions based on such incoherent presuppositions need not be answered because they cannot be answered. We need hardly apologize for our failure to do the impossible, for failing to answer questions of the form, "Why is (something self-contradictory) true?"

It is more plausible to suppose that our questioners are asking for an explanation of the Schrödinger equation in terms of some more general theory or principle. Such a theory or principle could not itself be a theory or principle of the same epistemological status as the Schrödinger equation itself. We are here presupposing that the Schrödinger equation or, alternatively, the fundamental quantum conditions (6.1) built into that equation, is the basic equation of quantum theory. It is an "explanation" of this equation that we are asking for. To explain this equation by employing some more basic physical relations is to violate the hypothesis under which we are supposed to be operating—that the Schrödinger equation is the most basic principle of the theory. Accordingly, the attempt to explain the Schrödinger equation by appeal to a more general theory or principle is self-defeating.[3]

The second of our cited questions requests, in effect, an "explanation" of (each of) Heisenberg's uncertainty relations, but Bohr and Heisenberg have given an explanation of these relations. Recall from chapter 1 that the uncertainty relations were there explained, in the sense that they were shown to be a deductive consequence of a wavelike description of physical systems, a description incorporating the fundamental quantum relations given in equation (6.1). Having explained the relation as a necessary consequence of the basic principles of quantum mechanics, we cannot, without embarking on a self-defeating exercise, attempt to explain these basic principles in terms of more basic ones.

The force of the above point can be brought out by further discussion of the now familiar multiple-slit experiment. We saw in chapter 5 that the anomalous interference patterns produced by beams of particles passing through and/or reflected from crystal lattices are readily explicable, given the quantum mechanical description of the incident beam. This wave functional description always refers, implicitly, to ensembles of similar particles, the position-momentum uncertainty relation placing a limit on the collimation of a beam of many simultaneously incident particles *or* on the reproducibility of the trajectory of a single-particle incident beam. Consider the case of a single-particle incident beam for a moment. Why is it that the trajectory of this particle is unreproducible and in fact indeterminable in advance but only retroactively? Consider what happens when our single-particle incident beam passes through a single slit or, equivalently for present purposes, a "pinhole." By making the dimension of the slit or pinhole very small we can fix the position of the particle as it passes through the aperture. Further, it must have some momentum or other the instant after passing through, and this momentum can be inferred retroactively after the particle hits a "screen" placed on the other side of the aperture from the source of the particles. But we have not determined the momentum the particle had before passing through since, in general, if the aperture is small enough the particle momentum will be disturbed upon passage, due to interaction with the rim of the aperture.

Hence we can hardly hope to reproduce the original trajectory since we do not know and could not know what the characteristic simultaneous position-momentum values of this original trajectory were. We could not know the (exact) original trajectory without, directly or indirectly, measuring it. But, and this is the essential and unavoidable quantum mechanical discovery, we also could not know it if we did try to measure it, for then we would be disturbing and changing, in an ultimately unpredictable way, the trajectory we were trying to measure; and anything we did succeed in "measuring" would not be the original trajectory of interest but rather some new trajectory produced by our measurement.

It is at this juncture that those wedded to classical intuitions will balk. Why is the measurement-produced disturbance ultimately unpredictable, they will ask? Why can we not infer the original trajectory from the produced one? Why, indeed? To infer the particle momentum before the measurement interaction from that possessed by the particle after the interaction, we would need to know, at least for some kinds of interactions, laws connecting earlier position-momentum values with later ones. That is, to suppose that the desired inference can be made successfully is to suppose that some mechanical processes are governed by exactly correct deterministic laws.

Notice what has been exposed. We began with the assumption that the uncertainty relations are true, and we sought an explanation of why these relations obtain, of why they are true. To "explain" the indeterminacy we were tempted to posit a set of deterministic laws, at least for some limited domain of processes. Such an "explanation" is incoherent for, far from explaining the universal applicability of the indeterminacy relations in question, it (implicitly) denies that the relations are universally applicable. We could consistently deny the truth of the relations and seek an explanation of why they appear to be true even though they are not—this is the approach of the hidden variable theorists and we saw in chapter 4 that their position is logically consistent even if incredible. What we cannot consistently do is accept the indeterminacy relations as true and then explain in deterministic fashion why the indeterminacy occurs. To attempt to do so is to attempt a deterministic explanation of the truth of indeterminism and so to embark on another self-defeating exercise.[4]

We are now in a position to answer the question concerning the seemingly nonstandard way of manipulating quantum mechanical probability distributions, illustrating the points under discussion by reference to diffraction experiments. The usual argument to establish the nonstandard character of quantum mechanical probability calculations runs as follows.[5] For a symmetrically arranged apparatus, with incident beam traveling in the $+x$ direction, the two-dimensional wave function representing the beam just after passage through the double slit is assumed to be of the form

$$\psi_{t=0}(x,y,t) = \psi(x,y,0) = (1/\sqrt{2})$$
$$\psi_1(x,y,0) + (1/\sqrt{2})\,\psi_2(x,y,0), \tag{6.2}$$

where ψ_1 and ψ_2 are the wave functions representing beams just after passage through slits 1 and 2, respectively. In particular

$$\psi_1(x,y,0) = \frac{1}{\delta\pi^{1/2}} \exp\left[-\frac{x^2 + (y-a/2)^2}{2\delta^2}\right] \exp\left(2\pi i p^0 x/h\right)$$
$$\tag{6.3a}$$

$$\psi_2(x,y,0) = \frac{1}{\delta\pi^{1/2}} \exp\left[-\frac{x^2 + (y+a/2)^2}{2\delta^2}\right] \exp\left(2\pi i p^0 x/h\right)$$
$$\tag{6.3b}$$

for particles of initial momentum p^0, slit widths δ, and slit spacing a. (Actually, the conceptual points at issue do not depend crucially on the details of these functional representations.) That is, the "state" of the beam after passage through the double slit is assumed to be a (linear) superposition of the states which occur when beams are passed through single slits.[6]

We are ultimately interested in the state of the beam or, equivalently, the form of the wave function at a screen located a distance, say, X from the slits. To calculate this function we make use of the fact that the particles of the beam do not interact with each other or, more precisely, that the probability of such interaction is negligible compared to the interaction between the beam and the slits. The accuracy of this assumption can always be guaranteed by using beams of sufficiently weak intensity. We further assume that the overlap between the wave "packets" represented by ψ_1 and ψ_2

is negligible, a good assumption provided $a \gg \delta$, i.e., provided the slit separation is large compared to the slit widths. Given these assumptions, we can calculate the form of the (total) wave function ψ at the screen at X by calculating separately the time dependence of ψ_1 and ψ_2 and superposing. That is, we can easily calculate

$$\psi(X,y,t) = (1/\sqrt{2})\, \psi_1\,(X,y,t) + (1\sqrt{2})\psi_2\,(X,y,t) \quad (6.4)$$

where $t = (m/p^0)\, X$ for particles of mass m. In calculating the time dependence, we are implicitly calculating the variation with respect to x at a given time since the beam moves in the $+x$ direction as time elapses.

Given the wave functional form (6.4), the probability that particles will strike the screen at location (X,y) is given by

$$P = P\,(X,y,t) = |\psi(X,y,t)|^2$$
$$= (1/\sqrt{2})^2\, |\psi_1(X,y,t) + \psi_2\,(X,y,t)|^2$$
$$= 1/2\, \{|\psi_1(X,y,t)|^2 + |\psi_2(X,y,t)|^2 + \psi_1^*\,(X,y,t)\psi_2(X,y,t)$$
$$+ \psi_2^*(X,y,t)\psi_1(X,y,t)\}.$$

But the probabilities of particles which are in states 1 and 2 striking the screen at location (X,y) are given respectively, by

$$P_1 = |(1/\sqrt{2})\psi_1(X,y,t)|^2$$

and

$$P_2 = |(1/\sqrt{2})\psi_2(X,y,t)|^2,$$

so that

$$P \neq P_1 + P_2$$

and the correct expression for P contains "interference" terms, $\psi_1^*\psi_2$ and $\psi_2^*\psi_1$, reminiscent of classical wave theories.

It is popularly alleged, on the basis of the foregoing analysis, that quantum mechanical probability calculations are "nonclassical" or "nonstandard" because it appears that the probability of particles arriving at a given location on the screen is not the sum of the probabilities of arrival from each of the individual slits, even though these individual probabilities are (correctly) assumed to be independent. This seemingly paradoxical result also has inspired proposals for revisions, such as those discussed in chapter 5, of classical logic in order to remove the paradox. But the correct procedure for removing the paradox has already been given in chapter 5. That analysis revealed that if we represent the incident beam by a "statistically"—i.e., indeterministically—interpreted wave function, then a quasi-periodic arrangement of material such as multiple-slit will not transmit or reflect beams with a continuous range of momentum values but rather beams with discrete values. A multiple-slit arrangement does *not* produce a result equivalent to the sum or (linear) superposition of the results produced by individual slits acting in isolation.

Thus the starting point of the argument sketched above is incorrect; accordingly, it is now seen to be completely unsurprising that a second misinterpretation, involving probabilities, is required to cancel out the original mistake and arrive at the correct result. If we analyze the problem correctly, we realize that each particle of the beam goes through one (and of course only one) of the two (or more) slits, but for the range of dimensions under discussion we cannot predict which one. The momentum of each of the transmitted particles does not range over continuous values as it would for a single slit but rather over discrete values; the probability of the particles striking the screen at location y is not a continuous function of y but rather is a correspondingly discrete function. Hence the total probability of arrival at screen coordinate y *is* the sum of the probability of arrivals at this point for particles passed through slit 1 plus the probability of arrival at this point for particles passed through slit 2, etc.

But here is the rub. We could not possibly verify by direct experiment the claim that the standard probability relations hold for *this* phenomenon. For any apparatus which determines which slit an individual particle passes through will effectively de-couple the individual slit from the neighboring slit; the multiple-slit plus additional apparatus will not act as a quasi-periodic arrangement of material but will produce the same result as the sum of individual slits acting in isolation. For such an experimental arrangement no characteristic interference lines will appear on the screen; rather, a pattern with a symmetrically located, single maximum appears.

Notice what has happened now. If we prepare an experimental arrangement which guarantees the accuracy of original assumption about the superposition or, equivalently, independence of the action of the individual slits then the standard probability relations—e.g., in the above calculation—hold. On the other hand, in the original experimental arrangement where the standard relations superficially appear to fail, a correct analysis of the situation reveals that it is the assumption about the independence (superposition) of the slit actions which fails, and it is only if we persist in retaining this inaccurate assumption that we are forced to (erroneously) interpret the probability relations in a nonstandard way in order to cancel out the first error and obtain the correct result.

One way of determining which slit a given particle passes through is to place a small, symmetrically located "mirror" in the plane perpendicular to the plane of the slits and much closer to the slits than the screen. If this "mirror" is not rigidly fixed but allowed to recoil when struck by particles, then from the direction of recoil we can infer which slit the particular particle passed through. Just as before, the multiple slit will pass particles with a discrete momentum spectrum but after collision with the mirror these momenta may take on any values; hence the pattern formed on the screen is that due to particles with a continuous spectrum and will not be an interference pattern. Readers in desperate need of imaginative exercise may attempt to envision an experimental apparatus which enables them to detect which slit a given particle passes through but does not destroy the previously observed interference pattern. Such attempts will fail.

The above points illustrate the sound core of Bohr's notion of complementarity: some pairs of experimental set-ups are mutually exclusive, in that one of the pair precludes obtaining the information the other is designed to yield. Bohr's mistake was to suppose that, because we cannot get all the information at one stroke, none of it is, strictly speaking, there to be got. Note that the sound insight in "complementarity" has nothing whatever to do with wave/particle dualism.

The conclusions above do not depend on features peculiar to slit-diffraction phenomena. Readers can verify the generalizability of the conclusions by applying corresponding analyses to other quantum mechanical phenomena displaying "probability interference," e.g., the Stern-Gerlach effect, involving superposition of spin states.

So far I have presented answers to some questions frequently cited as troublesome to defenders of a particle interpretation. I will now compare my interpretation, and the answers based on it, with the standard interpretation. This comparative study will show that, while my interpretation might seem quite different in spirit from the standard interpretation, actually a slight change in emphasis is sufficient to lead us from the latter to the former. Also, I will mention the differences between my defense of particle interpretations and those of Born and Landé. In spite of these differences, my debt to those two authors should already be apparent.

The problem of seemingly nonstandard probability relations provides a convenient starting point for comparison. Recall that Bohr and Heisenberg claim that we can never attribute exact intrinsic properties to physical systems. If we do attribute intrinsic properties to such systems, the attribution is at best an approximate one—i.e., the attributed properties necessarily possess a minimal and uneliminable indefiniteness and unreproducibility. Conversely, the only exact properties which can legitimately be attributed to physical systems are relational ones—those which are the outcome of interactions with other systems, including the especially important, for present purposes, measurement interactions. Given this general thesis, it is natural to interpret quantum mechanical probability statements as referring not to distribution of intrinsic properties over ensembles of particles but to the distribution of measurement results themselves. Thus, in the two-slit diffraction experiment, the sentence 'the probability that a particle in state

$$\psi(X,y,t) = (1/\sqrt{2})\,\psi_1\,(X,y,t) + (1/\sqrt{2})\,\psi_2(X,y,t)$$

has coordinate y is P' should be read as elliptical for the statement expressed by the sentence 'the probability that a position measurement performed on a particle in state

$$\psi(X,y,t) = (1/\sqrt{2})\,\psi_1(X,y,t) + (1/\sqrt{2})\,\psi_2(X,y,t)$$

will yield coordinate y is P'. This rendering of probability statements is held by advocates of the standard interpretation to be necessary in order to avoid paradox. That is, if we take the wave function $\psi(X,y,t)$ to refer to an ensemble of particles half of which are in state $\psi_1(X,y,t)$ and half in state $\psi_2(X,y,t)$, we will be led to the following erroneous conclusion. We will naturally suppose that the probability of finding a particle in state ψ_1 is (proportional to) $|\psi_1|^2$ and of finding a particle in state ψ_2, $|\psi_2|^2$. Thus we will suppose that since there are only two distinct states involved the total probability of finding a particle at (X,y) is the sum of the probabilities of finding particles in state 1 and 2, respectively, at this loctaion. That is, we will assume

$$P[\psi(X,y,t)] = P_1[\psi_1(X,y,t)] + P_2[\psi_2(X,y,t)]$$
$$= 1/2 \{ [|\psi_1(X,y,t)|^2 + |\psi_2(X,y,t)|^2 \},$$

which we know to be incorrect. This erroneous conclusion is avoided by the above-described move of the standard interpretation—before the position measurements the particles are allegedly not partly in state 1, and partly in state 2; they are in neither but rather are in the mixed or indefinite state

$$\psi(X,y,t) = (1/\sqrt{2}) \, \psi_1(X,y,t) + (1/\sqrt{2}) \, \psi_2(X,y,t).$$

Accordingly, the only probability which could possibly be involved here is the probability of finding particles in state ψ—i.e., the total probability of finding particles at X with coordinate y—and this probability is proportional to $|\psi|^2$, whose expansion contains the required interference terms. We could, of course, measure the relative number of particles in states ψ_1 and ψ_2 just after passage through the double slit. More precisely, we could ascertain that the probabilities of finding particles at location $(0, \pm a/2)$ are given, respectively, by

$$P_{1,2}(0, \pm a/2) = 1/2 \, |\psi_{1,2}(0, \pm a/2, 0)|^2.$$

But according to the standard interpretation this does not mean that half the particles were in state 1 and half in state 2 before these measurements. On the contrary, the measurement interactions caused the particles which were previously in the indefinite state ψ to go into either state ψ_1 or ψ_2.

The above description of the standard quantum mechanical interpretation of probability statements enables us to see two things. First, the interpretation is gratuitous in that it is only needed if we ignore the correct and well-known analyses of the phenomena in question. Second, there is a sense in which it reveals a correct and important point, but this sense is obscured by the misleading and ultimately counterintuitive nature of the interpretation. Take the second point first. There is a sense in which the measurement of the relative probabilities of states ψ_1 and ψ_2 does bring about or create these states, states which previously did not exist. That is, the "mirror" device described above does convert the beam transmitted through the double slit into a beam composed equally of ψ_1

state particles and ψ_2 state particles; furthermore, the transmitted beam was not so composed before striking the mirror and would not have been so composed when striking the screen had not the mirror intervened. This is the truth contained in the standard interpretation. But notice what the interpretation fails to reveal. Each particle passes through one and only one slit, though which slit is unpredictable. Further, the total probability of a particle strike at screen location (X,y) is the sum of probabilities of striking (X,y) when starting from each of the two slits; this is true whether the mirror apparatus is inserted or not.

The fact that a given particle goes through one and only one slit is not a fact created by the measurement involving the mirror. Equivalently, each particle has definite position and momentum at all times; these values are not created by measurements in the sense that prior to measurement it is senseless to suppose that they have such intrinsic properties. But the values of some of these variables are always altered by measurement interactions and, in fact, all interactions with other systems. So we cannot, in general, suppose that measured values correspond to those possessed before measurement. Thus we see that the slogan "measurement creates or brings about the measured states or values" is, strictly speaking, correct since a literal interpretation of it amounts merely to the claim that the measured values—as inferred from the behavior after the measurement interaction—are not in general those possessed before measurement. To move from this sound claim to the counterintuitive claim that the observed system therefore had no intrinsic properties before measurement is to be guilty of a non sequitur.

At just this juncture we must drive home a crucial point. Consider the objection—adumbrated by Mehlberg in reply to an early version of Landé's particle interpretation (both in Feigl and Maxwell 1961)—that attribution of intrinsic properties to ensemble members is gratuitous since what is calculated and "measured," namely, the (square of the) wave function, is not a quantity reducible to sets of intrinsic properties. The mistake here is by now depressingly familiar. Indeterministic particle theorists are being told they should be able to analyze the wave function (whose earlier and later values are connected, via Schrödinger's equation or the equivalent, in one-one, deterministic fashion) as a configuration of intrinsic properties. It is no criticism at all to point out that indeterministic interpretations are indeterministic, not deterministic. The positive content of indeterministic particle interpretations is that classically specified forces determine not exact future states but rather, via the wave function solutions to the wave equation, the relative long-run frequency of such states.

Left undisturbed, stable physical systems do not change their dynamical properties—energy, momentum, etc. When disturbed, as for example in measurement interactions, the disturbing forces *cause* whatever dynamical changes occur; they *determine* only that the changes will be within a certain specified range, discrete or continuous, as delimited by conservation laws, symmetry considerations, etc. Given that the equation we use has single-valued time-dependent solutions and that the coefficients—the Hamiltonian operators, etc.—specify exactly the (hypothetical) forces or potentials involved, where should we put the indeterministic aspect of the description of physical processes? There is only one place to put it: in the wave function; if it doesn't get in the description that way, it doesn't get in at all.

The notion of measurement interaction brings out another sound insight of the standard interpretation. Bohr and Heisenberg repeatedly remind us that interaction with another system disturbs the given system in "an unpredictable way." The correct interpretation here is that the interaction causes but does not determine the changes in the intrinsic properties of the two systems involved. Unfortunately, Bohr and Heisenberg do not want their sound insight formulated this way. They speak of a mysterious reduction of the "wave-packet" or "state-vector" upon interaction and imply that the existence of this process precludes attribution of intrinsic properties.

The attitude of many defenders of the standard interpretation is typically expressed by Heisenberg, who says that if no measurement —and presumably no calculation—could ever tell us what the intrinsic properties of a physical system were before measurement, e.g., what its simultaneous position and momentum were, then

It is a matter of personal belief whether . . . the past history of the electron can be ascribed any physical reality or not (Heisenberg 1930).

In the passage immediately preceding the quoted one, Heisenberg argues that such ascriptions are purely "speculative" since they can never be used as initial conditions in calculations of the future development of the system in question. Heisenberg, Bohr, et al. are concerned to defend the basic character and the legitimacy of the wave functional description of physical systems. They are anxious to avoid the indictment, brought by true believers in determinism such as Einstein and Bohm, that the standard quantum mechanical description of physical systems is ultimately "incomplete" and that newer, more refined deterministic theories of these systems and their processes are to be expected, just because of this incompleteness. Defenders of the standard interpretation fear that if they admit that physical systems have exact intrinsic properties— e.g., simultaneous position and momentum—then they will be committed to the admission that the standard theory is incomplete. We know that such fears are groundless.

Those who charge the standard theory with "incompleteness" have added nothing to the conversation beyond saying that the theory is indeterministic. Because the claim that position-momentum states at various times are connectable via deterministic laws is independent of the claim that physical systems have such simultaneous position and momentum properties, the claim that the theory must be incomplete because of the existence of such intrinsic properties is a non sequitur. The pejorative term 'incomplete' is appropriate only if we presume that some deterministic set of laws is true. The arguments of chapter 4 should have enabled us to wean ourselves away from this presumption.

The upshot of the preceding points is obvious. Once a defender of the standard interpretation realizes that fears about being committed to the admission that the standard theory is incomplete are groundless, there is nothing to prevent him from accepting the interpretation we have defended here. Both our interpretation and the standard interpretation are radically indeterministic interpretations; and it is only groundless fears that acceptance of the former interpretation will leave them open to attack by the common enemy —advocates of determinism—that prevents defenders of the latter interpretation from such acceptance. One way in which the fear in question is exhibited by defenders of the standard interpretation is in their frequent insistence that the physical state described by a quantum mechanical wave function is an "ultimate" or "pure" state and hence that "it is not appropriate to attempt to ascribe the uncertainty relations to something more fundamental" (Tomonaga 1966).

Clearly, this way of expressing the matter reveals that such defenders suppose that if we interpret quantum mechanical wave functions as referring to ensembles of similar systems, where each member of the ensemble has exact intrinsic properties such as simultaneous position and momentum, then we are leaving our theory open to the charge of "incompleteness." Accordingly, they reason, we must regard each of the systems in question as being in the same "pure" or "ultimate" state, as having the same properties —namely, as having no intrinsic properties but only relational ones such that relative to certain measurement interactions the system will yield a certain spectrum of measured results. Once again the gratuitous character of this move is obvious. Given the ultimately indeterministic character of all physical interactions, a theory can hardly be denigrated for admitting the existence of intrinsic properties such as simultaneous position and momentum and yet "failing" —as though success were not ex hypothesi precluded—to connect, say, earlier and later position-momentum pairs.

It will be instructive here to examine Einstein's attitude toward quantum theory. From the time of the 1927 Solvay Congress until his death in 1955, Einstein consistently branded quantum theory "incomplete." In a 1935 article (Einstein, Podolsky, and Rosen) he claimed to have shown this incompleteness, by means of a since-famous thought experiment. The cited article shows that if we accept as a sufficient condition for "physical reality"—i.e., what we are calling the existence of an intrinsic property—that an exact value of the property can be inferred without disturbing the measured system, then a certain thought experiment shows that simultaneous position and momentum values exist. Hence, since the quantum mechanical description, the wave function-Hamiltonian formalism, cannot in general predict these simultaneous values, quantum mechanics is an incomplete theory.

In his prompt reply, Bohr (1935) objected to Einstein's criterion for intrinsicality, saying, in effect, that only those measured properties whose values are connectable to their values at later times are real. Bohr brings out clearly the physics of the thought experiment:[7] two particles are passed through different slits in an originally stationary double-slit arrangement; by measuring the vertical component of momentum-recoil of the slit arrangement, we know the *sum* of the vertical momenta of the two particles and from the slit spacing we know the *difference* in the vertical positions of the two particles; hence measuring the position (or momentum) of either particle will allow us to infer exactly the position (or momentum) of the other. From this description of the situation, Einstein concludes that both particles have simultaneous position and momentum, since we could infer, after measurement on the other particle, the value of either property without affecting the particle in question. Bohr rejects this conclusion, because in measuring the position (momentum) of the first particle we disturb its momentum (position) and so lose all chance of inferring the momentum (position) of either of the particles.

Two aspects of this Bohr-Einstein dispute are important for our present purposes. First, the conclusion Einstein established by his thought experiment is the same conclusion which can be established, if at all, by the much simpler "pinhole" experiment I have relied on several times in earlier chapters. This simpler thought experiment was well known to all parties concerned—Bohr, Heisenberg, Born, Einstein, etc.—in the late twenties. Second, Bohr is saying, in effect, that the existence of intrinsic properties presupposes that changes in such properties are deterministically describable, a line of argument disposed of in chapter 2. Conversely, Einstein is complaining that quantum mechanics is incomplete because it is indeterministic.

It would be a mistake to accept the overly generous assessment of Born and Pauli, that Einstein was *only* concerned with showing the existence of intrinsic properties and was willing to accept the indeterministic quantum theory as an ultimately satisfactory description of physical processes.[8] Einstein himself sometimes seemed to make such claims, but other passages in his writings belie the claims; at best, his attitude was consistently ambivalent. In the 1935 article he told us that only the concepts of a *complete* theory are satisfactory and that he believed such a complete theory could be discovered (Einstein, Podolsky, and Rosen, pp. 777, 780). In 1944 he was still chiding Born for believing "in the God who throws dice [while] I [believe] in complete law and order . . . even the great initial success of the quantum theory does not make me believe in the fundamental dice-game" (Born 1971, p. 149). In 1947 he told us "I cannot believe in [the indeterministic approach] because the [quantum] theory cannot be reconciled with the idea that physics should represent [objective, intrinsic situations]" (Born 1971, p. 158). In 1948 he believed that, come what may, "it is expedient in the end to keep the basis of the theory free from statistical concepts." Later in the same year we find him saying "I am therefore inclined to believe that . . . quantum mechanics [properly interpreted] has to be regarded as an incomplete and indirect description of reality, to be replaced at some later date by a more complete and direct one" (Born 1971, pp. 163, 173). In 1953 he still did *not* believe that "there is no . . . complete law for the complete description of [a] single system" (Born 1971, p. 209). Finally, in 1954, in the course of trying to convince Born of just the opposite conclusion, Pauli admits that Einstein still did not believe that, on *any* interpretation, quantum mechanics is a correct description of reality; it is unsatisfactory because incomplete, i.e., indeterministic (Born 1971, p. 226).

The interpretation we have been advocating in this essay is hardly original; in a way it is even older than the standard Copenhagen interpretation. The spirit of our indeterministic particle interpretation in its basic essentials was suggested by Born as early as 1926. In particular, the crucial insight that the central idea of classical mechanics is separable into two distinct ideas is due to Born, yet he frequently claimed to accept the Copenhagen interpretation of Bohr and Heisenberg: a paradoxical state of affairs. Perhaps the most accurate way of putting the matter is to say that although the cited separability of the central idea of classical mechanics is an immediate consequence of some of Born's ideas, Born himself never actually took the step of drawing this consequence; instead, influenced by the same groundless fears which have restricted advocates of the standard interpretation, Born concluded that application of the term 'particle' to quantum mechanical systems involved a change in meaning compared to the classical usage. Thus, Born realized immediately that in quantum mechanics

forces do not, as they did classically, cause accelerations of the particles; they have no direct bearing on the motion of the particles. . . . the forces determining the vibrations of a certain function ψ that depends on the positions of all the particles . . . and determine them because the coefficients of the differential equation for ψ involve the forces themselves (Born 1956, p. 10).

Here we have ingredients sufficient to suggest the conclusion that bodies travel in definite paths, i.e., that they are particles, even though the relevant forces do not determine these paths uniquely but only determine the relative probability of various paths. But Born did not draw this conclusion, even at this stage of his thinking. Instead he held that although

it is not forbidden to believe in the existence of simultaneous position-momentum coordinates . . . they will only be of physical significance when methods have been devised for their experimental observation (1956, p. 11).

Thus we see that Born at this stage was agnostic about the existence of simultaneous position and momentum values, and the above passages were written just before Heisenberg's discovery of the uncertainty relations. Once that discovery was made, Born felt that it precluded the assertion that such simultaneous values exist. Unlike other adherents of the Copenhagen interpretation, he seldom if ever went so far as to say that such an assertion is "senseless." Rather, his attitude was that the uncertainty relations supported and indeed demanded the now familiar conclusion that the only exact properties attributable to objects are relational ones. Just as did Bohr and Heisenberg, Born felt this conclusion was necessary to avoid any reinstatement of determinism. Specifically, Born felt that the truth of the uncertainty relations implied that quantum mechanical objects are items "with which the concepts of position and momentum cannot be associated in the usual way" (Born 1956, p. 187). Obviously, if such an implication holds we cannot make the usual association without denying the uncertainty relations, a denial completely unacceptable to Born. But the conclusion of the cited implication is ambiguous, and we have seen that the implication holds only in the sense in which "usual way" includes both the attribution of simultaneous position-momentum pairs *and* the further claim that deterministic laws connect earlier and later pairs. Had Born held the separability of the "usual way" firmly in mind, he could have accepted our interpretation and still avoided the dreaded reinstatement of determinism.

Born is not the only precursor of the interpretation herein defended. Landé has voluminously defended an indeterministic particle interpretation of quantum mechanics (1955, 1960, 1965). In particular, Landé, more than any previous writer, realized the significance of the analysis of the crystal lattice diffraction experiments presented in chapter 5. Yet, Landé has not been content with interpreting the standard theory as given but has attempted to deduce certain fundamental aspects of the theory from principles allegedly more general, such as a "cause-effect continuity" principle.

At best this explanatory exercise of Landé's is above and beyond the call of duty; and, in fact, somewhat confused.[9] Completely apart from this general confusion, the details of Landé's analysis reveal that even he has not sufficiently freed himself from some of the subtler restraints of the standard interpretation. Landé feels obliged to offer an explanation of why probabilities "interfere," rather than add, in certain mechanical phenomena. To effect this explanation he relies on other general principles, involving reflexivity, symmetry, and transitivity. Even if his analyses are formally correct, this exercise is misguided, for it obscures rather than reveals the important insight that probabilities in fact never do interfere. Landé fails to grasp that the same indeterministic particle analysis of, e.g., the crystal lattice diffraction experiments, which obviates any realist wave interpretation or revision of classical logic also obviates, for exactly the same reasons, the admission of probability interference.

Apart from the yet untouched issues of field theory, we have now completed our survey of the philosophic problems arising in the interpretation of quantum mechanics. It would be redundant to review in detail our conclusions, but two central points are worth explicit mention. First, if we are willing to accept a single premise —that the correct description of physical processes is fundamentally indeterministic, not deterministic—then all the philosophic problems of interpretation in quantum mechanics become tractable. We can consistently carry out a particle interpretation of phenomena involving material objects of all sizes, from electrons to galaxies, without equivocating on the key term 'particle'; e.g., we can explain probability interference because we can come to realize that it never in fact occurs. Equally important, we can appreciate the misguided and superfluous character of various alternate solutions to the problems in question—revisions of logic, hidden variable theories, "dualism," etc. Second, the cited indeterministic premise is by no means an a priori assumption but is the only hypothesis supported by the present evidence.

7
Relativistic and Field Theory Considerations

Limitations of Nonrelativistic Quantum Theory

The theory we have been discussing so far in this study is a theory of limited scope. It works well for nonrelativistic interactions of material particles, but it is incorrect whenever relativistic speeds are involved; furthermore, it cannot handle those processes or subprocesses constituted by the absorption, emission, or transmission of electromagnetic radiation. In this final chapter investigations of relativistic formulations of quantum mechanics and the quantum theory of fields will enable us to establish two crucial points. First, there is nothing in the more generally correct theory which counts decisively against our particle interpretation of quantum mechanics, i.e., so far as material particles are concerned; indeed, some field theory considerations provide new support for our interpretation. Second, the widespread thesis that, when field theory considerations are taken into account, wave and particle interpretations can be shown to be entirely equivalent, the choice between them being one of situational convenience, will be exposed as completely unfounded. Even the alleged *formal* equivalence between the two pictures—wave and particle descriptions—will be shown to be oversimplified and misleading for most material particles, i.e., for "fermions." More important, even where the formal analogy applies exactly, it will be shown to ignore disanalogies which are more striking, suggestive, and important than the analogies. In the course of this discussion the important differences between material particles and photons, or "units of electromagnetic interaction," will be enumerated.

Recall the inherent limitations of any nonrelativistic theory, such as Schrödinger's quantum mechanics. First of all, the time-dependent Schrödinger equation does not transform properly; it is not invariant in form under the important transformations of coordinates known as the Lorentz transformations. A basic presupposition of the special theory of relativity is that the form of physical laws must be invariant under such Lorentz transformations, which is merely a precise way of saying they must be of the same form for all observers in inertial frames. So special relativity, and the vast experimental evidence confirming it, precludes our regarding the Schrödinger equation as generally correct.

That the Schrödinger equation does not transform properly is obvious by inspection, since the first time-derivative is involved but the space derivatives are second derivatives. Clearly, the source of this difficulty is the fact that the Schrödinger equation is based on the classical or nonrelativistic conservation of energy relation, e.g.,

$$E = \mathbf{p}^2/2m \qquad (7.1)$$

for a free particle. In the derivation of the Schrödinger equation, the energy and momentum variables of equation (7.1) are replaced by corresponding differential operators

$$E \longrightarrow i\hbar \, \frac{\partial}{\partial t}$$

$$\mathbf{p} \longrightarrow -i\hbar\nabla.$$

But then equation (7.1) will obviously yield an equation without the symmetry—between space and time variables—required for a relativistically correct equation. To remove the source of the difficulty, we must use, explicitly or implicitly, the relativistic energy equation

$$E^2 = m^2 \, c^4 + c^2 \, \mathbf{p}^2. \qquad (7.2)$$

Clearly, equation (7.2) provides the requisite symmetry between space and time variables. Equation (7.2) also reminds us of another relativistic consideration, related to a second limitation of the Schrödinger equation. The total energy of a system of particles is not just the total kinetic or even kinetic plus potential energy of the system. Rather, the system possesses a quantity of "rest energy," a quantity proportional to the mass(es) involved. Energy can be created or destroyed by destroying or creating a corresponding quantity of mass. But if we wish to describe processes in which particles are destroyed—not broken up into subparticles but literally annihilated, leaving in their wake only energy; or created—not from other particles but from (electromagnetic) energy—then the Schrödinger equation could not be adequate for this new descriptive task. For the statistically interpreted Schrödinger equation presupposes conservation of particles; the probability of finding a given particle at various spatial locations varies from location to location, but the total probability of finding the particle somewhere in space must be equal to 1.0. If we allow creation and annihilation of particles, and direct experimental evidence as well as the theory of relativity demands that we do, this normalization condition breaks down.

There are two obvious corollaries, concerning photons, stemming from the above discussion. First, the Schrödinger equation could not possibly handle photons since the condition for adequacy of nonrelativistic theories

$$v^2/c^2 << 1$$

is never satisfied for photons, where $v = c$. Second, photons are characteristically created or destroyed in processes where the number of material particles is constant. Hence at best we could hope that the Schrödinger theory is adequate for the material particle part of these processes; it could not be adequate for the whole of such processes.

There is a third difficulty with the Schrödinger theory, not an inherent limitation but an embarrassment to theoreticians. The classical Hamiltonian or energy equation does not automatically include all the interactions known to be involved in material particle processes. In particular, a charged particle such as an electron will possess a magnetic moment associated with its spin and because of this moment will interact with externally applied magnetic fields. It is this interaction for which an extra term must be provided, seemingly on a completely ad hoc basis, in the classical Hamiltonian. It is one of the virtues of the Dirac formulation of relativistic quantum mechanics that the existence of this spin-related interaction becomes a natural consequence of the general principles of the theory.

Relativistic Formulations

The most straightforward way of obtaining a relativistically correct wave equation is to make the usual operator substitutions in equation (7.2). This procedure yields

$$(\Box^2 - M^2)\,\psi = 0, \tag{7.3}$$

where

$$\Box \equiv \nabla - \frac{1}{ic}\frac{\partial}{\partial t}, \quad M \equiv mc/\hbar.$$

Equation (7.3) is usually called the Klein-Gordon equation, though it was first proposed by Schrödinger. In interpreting the Klein-Gordon equation we are guided by our experience in interpreting the nonrelativistic Schrödinger equation, where it is natural to interpret $|\psi|^2$ as the particle probability density and

$$\mathbf{S} = -\frac{i\hbar}{2m}(\psi^*\nabla\psi - \psi\nabla\psi^*)$$

as the probability current density, i.e., the probability per unit time that a particle will cross a unit surface in a given direction. It is "natural" to do this for two reasons. First, the probability so defined is always positive since

$$|\psi|^2 = \psi^*\psi > 0.$$

Second, P and $\mathbf{S}$ satisfy the continuity equation

$$\frac{\partial P}{\partial t} + \nabla \cdot \mathbf{S} = 0. \tag{7.4}$$

Equation (7.4) is an immediate consequence of Schrödinger's equation, and application of Gauss's theorem to equation (7.4) shows that the integral over all space of P is a constant.

When we try to find the Klein-Gordon quantities representing probability density and current density, however, we encounter apparent difficulties. We know that under Lorentz transformation volume elements contract, i.e.,

$$d^3x \longrightarrow d^3x\,(1 - v^2/c^2)^{1/2}.$$

146

So if the integral over all space of the probability density is to remain constant, we must have

$$P \longrightarrow P(1 - v^2/c^2)^{1/2}.$$

That is, P must be the fourth or timelike component of a 4-vector density. Our relativistic continuity equation is

$$\frac{\partial}{\partial x_\mu} S_\mu = 0, \qquad (7.5)$$

where

$$S_\mu = (\mathbf{S}, icP).$$

It is easily shown that

$$S_\mu = (\text{const.}) \ (\psi^* \frac{\partial \psi}{\partial x_\mu} - \frac{\partial \psi^*}{\partial x_\mu} \psi),$$

where ψ is a solution of the Klein-Gordon equation, satisfies the continuity equation (7.5). Furthermore, in the nonrelativistic limit $(E \cong mc^2)$

$$\psi \cong \psi_s \exp -imc^2t/\hbar,$$

where ψ_s is the corresponding nonrelativistic Schrödinger equation solution and

$$S_0 = -iS_4 \cong (2imc/\hbar) \ (\text{const.}) \ |\psi_s|^2$$
$$\mathbf{S} = (\text{const.}) \ [\psi_s^* \nabla \psi_s - (\nabla \psi_s^*) \psi_s].$$

That is, the components of the Klein-Gordon current density correspond to the Schrödinger current density and probability density in the nonrelativistic limit. But we can hardly interpret

$$P = \frac{i\hbar}{2mc^2} \ (\psi^* \frac{\partial \psi}{\partial t} - \frac{\partial \psi^*}{\partial t} \psi)$$

as probability density, since P can take on negative as well as positive values. Ultimately, the source of this difficulty can be traced back to our original equation, (7.2), which is quadratic in energy. Theoretically, negative as well as positive energy values satisfy this equation, and wave functions corresponding to these negative energies are equally good solutions of the Klein-Gordon equation. This suggests that although S_μ cannot be interpreted as a probability current density, equation (7.3) might still be physically correct if S_μ is interpreted instead as a charge current density where the negative energy solutions are identified as the wave functions for a particle oppositely charged compared to the particle described by the positive energy solutions. This interpretation is just the sort we want in the end, since there are known processes in which it is the total charge rather than the total number of particles which is conserved.

Even if we make a virtue of necessity by shifting from a probability current interpretation to a charge current interpretation, the Klein-Gordon equation is unsatisfactory as a description of most "elementary" particles—e.g., electrons, protons, neutrons, and their corresponding antiparticles. That is, the Klein-Gordon equation reduces in the nonrelativistic limit to the Schrödinger equation without the above-mentioned spin-related interaction term. Since we know experimentally that a correct description must include this spin interaction term, the Klein-Gordon equation as it stands (7.3), is correct only for spinless particles and not for those with intrinsic spin. We could doctor up the Klein-Gordon equation in ad hoc fashion, just as Pauli doctored up the Schrödinger equation, so as to obtain the required spin term. There is little incentive for this procedure, however, when we can find a relativistic wave equation which automatically yields the correct spin term.

The original difficulty which inspired the search for a relativistic equation was the fact that space and time derivatives are of different order in the Schrödinger equation. The Klein-Gordon approach is to make both derivatives second order; the Dirac approach is to make them both first order. Dirac showed that the following equation is Lorentz invariant and consistent with the Klein-Gordon equation and hence with the correct Hamiltonian:

$$(\gamma_\mu \frac{\partial}{\partial x_\mu} + M)\psi = 0. \tag{7.6}$$

To show the latter claim, operate on equation (7.6) from the left with

$$\gamma_\lambda \frac{\partial}{\partial x_\lambda} - M$$

to get

$$(\gamma_\lambda \gamma_\mu \frac{\partial^2}{\partial x_\lambda \partial x_\mu} - M^2) \psi = 0,$$

which is identical with the Klein-Gordon equation, provided

$$\gamma^\lambda \gamma^\mu + \gamma^\mu \gamma^\lambda = 2\delta_{\lambda\mu}. \tag{7.7}$$

Clearly, the γ's cannot be numbers, real or complex, if they are to satisfy the anticommutation relations, (7.7). We can, however, take the γ's to be matrices; then ψ is to be regarded as a column matrix or column vector—an item with several components and hence capable of representing particles with extra degrees of freedom, such as intrinsic spin. One can find various sets of γ matrices satisfying equation (7.7), but a particularly convenient set is

$$\gamma_k = \begin{pmatrix} \phi & -i\sigma_k \\ i\sigma_k & \phi \end{pmatrix}, \quad k = 1, 2, 3,$$

$$\gamma_4 = \begin{pmatrix} I & \phi \\ \phi & -I \end{pmatrix},$$

where

$$I = \begin{pmatrix} 1 & 0 \\ 0 & 1 \end{pmatrix}, \quad \phi = \begin{pmatrix} 0 & 0 \\ 0 & 0 \end{pmatrix},$$

and the σ_k are the 2×2 Pauli spin matrices correctly describing spin properties in the nonrelativistic theory.

Using the formalism so far described, several results can be established. First, equation (7.6) is indeed Lorentz-invariant, just as claimed by Dirac. Second, if we write (7.6) in Hamiltonian form, we can see that the *total* angular momentum which is conserved for isolated systems is the orbital angular momentum plus a spin angular momentum of magnitude $\hbar/2$. So the Dirac formalism, unlike that of the Klein-Gordon equation, automatically predicts the existence and correct magnitude of the intrinsic spin observed for electrons, protons, neutrons, etc. A further result of this same analysis is that the magnetic moment of the electron in an external field is correctly predicted for the first time. In the nonrelativistic theory the predicted moment differed from the measured value by a factor of 2, and this factor had to be introduced into the theory on an ultimately ad hoc basis. Third, the Dirac formalism is tentatively amenable to a probabilistic interpretation analogous to that appropriate for the nonrelativistic Schrödinger equation. The quantities

$$S = i \, c\psi^\dagger \gamma_4 \gamma_\mu \psi$$

and

$$P = \psi^\dagger \psi$$

satisfy the continuity equation and the integral of P over all space is constant. Further, P is always a positive quantity. (The row vector $\psi^\dagger$ is the transpose of the column vector with the elements of $\psi^\dagger$ being the complex-conjugates of the elements of ψ.)

The Dirac theory described above has achieved great predictive, descriptive, and explanatory success, not only with regard to the cited phenomena but also with regard to phenomena involving the interaction of material particles with radiation—i.e., with electromagnetic fields. In spite of these happy features, it cannot be regarded as generally correct. The source of the difficulties with the Dirac theory can be traced back to the existence of solutions corresponding to negative energy values. There are four linearly independent solutions for the Dirac equation, corresponding to positive and negative energy solutions for each of the two independent spin states. To see the importance of these negative energy solutions and to see why they cannot simply be ignored, recall an important difference between quantum and classical mechanics: consider the process whereby atoms in excited states emit radiation and make transitions to states of lower energy. This process can go on until the atom reaches its ground state or state of lowest positive energy. If we include the negative energy states whose existence is implied by the Dirac equation, the usual "ground" state is not really the lowest energy state; on the contrary, there is a continuous range of negative energy states from $-mc^2$ to $-\infty$ for particles of mass m. What is to prevent electrons from making transitions from the ground state to one of these negative energy states by emitting a photon of energy $E \geqq E_{\text{ground}} + mc^2$? In classical physics such a transition would be impossible since systems cannot make transition between distinct energy levels without passing through each of a presupposed continuous set of intermediate energy levels. But discrete energy transitions or quantum jumps are quite possible in quantum mechanics; hence the negative energy solutions cannot simply be discarded. So the Dirac theory seems to predict transitions which are inconsistent with the observed stability of atomic energy ground states.

Dirac attempted to resolve this dilemma by the hypothesis that, under normal conditions, all the negative states are filled. If this is correct, then the undesirable transitions to negative energy states are precluded because of the Pauli exclusion principle: at most one electron (or proton or neutron) can occupy a given energy state at one time. However, the reverse process would still be allowed; i.e., a negative energy electron could absorb a high energy photon and make a transition to a positive energy state. Further, the "hole" in the Dirac sea of negative energy states left by this escalated particle appears to be positively charged—the charge of the sea appears to increase algebraically because of the loss of a negative charge. So, observationally, the escalation process appears as the creation of a *pair* of particles of positive energy—one the usual negatively charged electron, the other a positively charged electron. The existence of this positively charged electron or "positron" has since been confirmed by experimental evidence.

Notice again why this "hole" theory works; in particular, notice the relation to the tentative probabilistic particle interpretation of the Dirac equation sketched above. In effect the hole theory argues that, just as required by the probability interpretation, the total number of particles involved is conserved—all that happens in pair creation is that a negative energy electron is escalated into a positive energy state; similarly, in pair annihilation a positive energy electron drops into a negative energy state. In each case the hole theory maintains the fiction that no particles are created or destroyed; the appearance and disappearance of positrons is treated as the appearance and disappearance of the corresponding holes in the Dirac sea. It is the difference between the number of positive energy electrons and the number of positrons or holes which is treated as conserved. Whenever a new hole is created, a new positive energy electron is also created, so that the difference remains constant. Similarly, whenever a hole is filled, a positive energy electron disappears. So we are able to maintain a conservation-of-particles interpretation of the Dirac theory, but only at the price of introducing an infinity of new particles, the electrons filling all the negative energy states.

Clearly, there is something very suspicious about this procedure; in order to guarantee conservation of particles we are forced to introduce an infinity of new particles and so depart from the spirit, at least, of the conservation-of-particles idea. And in fact there are many elementary particle reactions where even the difference between the number of electrons (or protons or neutrons) and the number of holes does not seem to be conserved. Hence a generally satisfactory theory is one in which particle creation and annihilation is not tied to the appearance or disappearance of holes in the corresponding sea of negative energy particles. The quantized field theory of Dirac particles, discussed in the next section, is such a theory.

We might suppose that because of the failure of a straightforward particle density interpretation of the Dirac equation that such interpretations are precluded and that we are forced to adopt a field interpretation for the Dirac wave function. In general, $P = \psi^{\dagger}\psi$ must be interpreted as a charge density rather than a particle density; the question is whether we should take this as decisive evidence that the systems described by the Dirac equation are not particles but fields, not discrete, (approximately) localizable carriers of charge, mass, energy, and momentum but continuous distributions of charge, mass, energy, and momentum. Such a field interpretation is attractive with regard to the Dirac equation and even more attractive with regard to the Klein-Gordon equation, the equation for spinless systems, where a straightforward particle interpretation leads to anomalous negative probabilities. Parenthetically, note that relativistic wave equations appropriate for systems of any intrinsic spin value are derivable; the Dirac equation amounts to a factorization of the Klein-Gordon equation but this factorization is not unique—other factorizations corresponding to other spin values can be found. Hence we might claim that particle interpretations are precluded for any relativistic wave equation.[1] This conclusion, while natural, is too hasty.

Let us regard the various relativistic wave equations as equations governing the probability density of particles, equations which are correct in the absence of actual physical fields strong enough to permit pair production and to accelerate particles to the energies required for pair annihilation. The systems described by the various relativistic wave equations for various material particles do exhibit discrete properties; their masses and charges are well known to be quantized. Further, apart from, or, rather, in between such processes as pair production and pair annihilation, the spatio-temporal behavior of the systems corresponds to the continuous trajectories of localized carriers of energy and momentum.

Our attitude amounts to saying that once it is recognized that relativistic wave equations (before the quantization process to be discussed in the next section) are not generally correct—do not accurately describe all physical processes—then its lack of general adequacy cannot be used against any particular interpretation of the theory as applied to those more limited processes which it does accurately describe.[2] If we exclude any given interpretation, we should do so by pointing out that it will not work for the generally correct theory; we should not proceed by pointing out that it will not work for a theory which all parties concede is inadequate anyway. Of course, to say that the (unquantized) relativistic wave equations are not adequate for pair creation, pair annihilation, etc., is to make a severe indictment of these theories, for it amounts to saying that they are relativistic theories which cannot handle those processes most characteristic of relativity—high energies, conversion of mass to energy and vice versa, etc.

Field Theory

The claim I wish to examine, very critically, in this section is the claim that within the most generally correct description of physical processes—quantum field theory—there is a perfect analogy between the behavior of material particles and the behavior of electromagnetic radiation.

Classically, electromagnetic phenomena are described by Maxwell's equations,

$$\nabla \cdot \mathbf{E} = \rho$$

$$\nabla \times \mathbf{B} - \frac{1}{c}\frac{\partial \mathbf{E}}{\partial t} = \mathbf{j}/c$$

$$\nabla \cdot \mathbf{B} = 0$$

$$\nabla \times \mathbf{E} + \frac{1}{c}\frac{\partial \mathbf{B}}{\partial t} = 0,$$

a set of equations relating electric ($\mathbf{E}$) and magnetic ($\mathbf{B}$) field vectors and source strengths—i.e., charge distribution (ρ)—and charge current ($\mathbf{j}$). It is convenient to introduce a field matrix F, a vector potential A_μ, and a four-vector charge current j_μ:

$$F = \begin{Bmatrix} 0 & B_3 & -B_2 & -iE_1 \\ -B_3 & 0 & B_1 & -iE_2 \\ B_2 & -B_1 & 0 & -iE_3 \\ iE_1 & iE_2 & iE_3 & 0 \end{Bmatrix}$$

$$\frac{\partial A_\nu}{\partial X_\mu} - \frac{\partial A_\mu}{\partial X_\nu} = F_{\mu\nu}$$

$$j_\mu = (\mathbf{j}, ic\rho).$$

With these introductions the (four-dimensional) charge current is conserved and the last pair of Maxwell's equations is automatically satisfied. Experimentally, we know that it is the field quantities, the components of the vectors $\mathbf{E}$ and $\mathbf{B}$, which are measurable; the potential A_μ is to be regarded as a mathematically convenient introduction. Accordingly, we can always choose the form of A_μ so that the first pair of Maxwell's equations assume the simple form

$$\square \, A_\mu = -j_\mu/c.$$

Furthermore, we can always choose A_μ so that $\mathbf{A}$, or at least the component of $\mathbf{A}$ perpendicular to the direction of propagation of a disturbance radiated throughout the fields, satisfies

$$\nabla \cdot \mathbf{A} = 0 \ (\text{or } \nabla \cdot \mathbf{A}_\perp = 0 \text{ where } \mathbf{A} = \mathbf{A}_\perp + \mathbf{A}_{||}).$$

With this choice we have a formalism appropriate for the description of electromagnetic waves—light waves, radio or television waves, etc.—which are known to consist of oscillations transverse or perpendicular to their direction of propagation. For such a radiation field, in the absence of sources, A_μ satisfies the wave equation

$$\square^2 A_\mu = 0.$$

Still speaking within the realm of classical description, we know that the radiation field described by the vector potential $\mathbf{A}$ can also be represented as an infinite collection of independent harmonic oscillators. That is, we can always expand $\mathbf{A}$ in a Fourier series

$$\mathbf{A}(\mathbf{x}, t) = (\text{const.}) \, \Sigma_{\mathbf{k}} \, \sum_{\alpha=1}^{2} [c_{\mathbf{k}, \, \alpha} \, (t) \, \epsilon_{\mathbf{k}}^{\alpha} \exp (i\mathbf{k} \cdot \mathbf{x}) + c^*_{\mathbf{k}, \, \alpha}$$

$$(t) \, \epsilon_{\mathbf{k}}^{\alpha} \exp (-i\mathbf{k} \cdot \mathbf{x})]$$

where the polarization vectors ϵ are perpendicular to each other and to the direction of propagation, given by $\mathbf{k}$; in general, components corresponding to all possible directions are possible—the field has a potentially infinite number of degrees of freedom at every space-time location $(\mathbf{x}, t)$. If the time dependence of the coefficients is represented as is the time dependence of harmonic oscillators, namely,

$$c_{\mathbf{k},\alpha}(t) = c_{\mathbf{k},\alpha}(0) \exp\left(-ic|\mathbf{k}|t\right)$$

$$c^*_{\mathbf{k},\alpha}(t) = c^*_{\mathbf{k},\alpha}(0) \exp\left(+ic|\mathbf{k}|t\right)$$

then $\mathbf{A}(\mathbf{x},t)$ is real and satisfies the wave equation. Using periodic boundary conditions on the product functions

$$\epsilon_{\mathbf{k}}^{\alpha} \exp\left(\pm i\mathbf{k}\cdot\mathbf{x}\right),$$

we can show that the total energy of the radiation field is given by

$$H = \Sigma_{\mathbf{k}} \sum_{\alpha=1}^{2} 2|\mathbf{k}|^2 c^*_{\mathbf{k},\alpha} c_{\mathbf{k},\alpha}.$$

If we now define the quantities

$$Q_{\mathbf{k},\alpha} = \frac{1}{c}\left(c_{\mathbf{k},\alpha} + c^*_{\mathbf{k},\alpha}\right)$$

$$P_{\mathbf{k},\alpha} = -i|\mathbf{k}|\left(c_{\mathbf{k},\alpha} - c^*_{\mathbf{k},\alpha}\right),$$

then

$$H = \Sigma_{\mathbf{k}} \sum_{\alpha=1}^{2} 1/2 \left(P_{\mathbf{k},\alpha}^2 + c^2|\mathbf{k}|Q_{\mathbf{k},\alpha}\right)$$

and

$P_{\mathbf{k},\alpha}$ and $Q_{\mathbf{k},\alpha}$ satisfy the familiar equations for the canonical variables—i.e., generalized coordinates and generalized momenta—of a mechanical system:

$$\frac{\partial H}{\partial Q_{\mathbf{k},\alpha}} = -P_{\mathbf{k},\alpha} \quad \text{and} \quad \frac{\partial H}{\partial P_{\mathbf{k},\alpha}} = +Q_{\mathbf{k},\alpha}.$$

So we have constructed a formal analogy between the description of a radiation field or electromagnetic wave-field and the description of an infinite collection of independent mechanical oscillators. At this juncture, before making the transition to quantum theory, it is important to emphasize the formal character of the constructed analogy. What we have done is represent a field oscillation at a particular spatial location by a superposition of orthogonal modes of oscillation, each mode of oscillation satisfying familiar mechanical oscillator equations. But these (plane wave) modes of oscillation do not correspond to mechanical oscillations—the items which are oscillating are not material particles, they possess no mass; as is well known, the oscillations can occur in regions evacuated of all material bodies.

Hence the oscillations in question are not the oscillations of the particles of a medium supporting the oscillations—supporting and transmitting the electromagnetic waves—since in general there is no such medium. Nor does the transmitted radiation itself consist of oscillating "particles" that travel through space in the direction of propagation. It is the wave, the disturbance, the field oscillation, which travels through space. The fictitious, massless mechanical oscillator "particles" remain in place. They do not move in the direction of propagation, and their oscillations are not spatial displacements in any direction, whether perpendicular to or along this propagation direction. What "oscillates" is the relative strength or weighting factor for each term of the plane wave Fourier series expansion. These simple points will be important in the sequel, when we examine the quantum theory of radiation.

The basic idea of the quantum theory of radiation is that, contrary to classical presumption, electromagnetic waves of given wavelength (or frequency) are not free to take on any values of momentum (or energy). Rather, waves of frequency v or length λ should be regarded as always carrying energy $E = hv$ and momenta $p = h/\lambda$. In terms of the cited formal analogy we can express this idea by saying that the behavior of waves of frequency v is formally analogous to the behavior of a collection of "photons" or massless "particles," each of energy nhv, with n taking only integer values. With these quantum mechanical restrictions, Planck and Einstein were able to explain, respectively, the blackbody radiation spectrum and the photoelectric effect.

In the quantum theory of radiation we want a formalism in which the cited restrictions are built in. To achieve this end, it is sufficient to make a move analogous to that made in nonrelativistic quantum theory. That is, just as we replaced the dynamical variables—energy, momentum, etc.—of mechanical systems by operators, in the nonrelativistic Schrödinger theory, so, in the quantum theory of radiation, we replace the generalized "coordinates" and "momenta" of the (massless) radiation oscillators by operators satisfying various commutation relations. (This replacement of real-number valued variables by mathematical objects functioning as operators—e.g., matrices, differential or integral operators, etc.— is the process usually referred to as 'quantization' although this term is sometimes used to refer to the results of making the replacement rather than to the replacement itself.) In particular, the operators for the radiation oscillators are assumed to satisfy the commutation relations

$$Q_{\mathbf{k},\alpha} P_{\mathbf{k}',\alpha'} - P_{\mathbf{k}',\alpha'} Q_{\mathbf{k},\alpha} = i\hbar \delta_{\mathbf{k}\mathbf{k}'}\,\delta_{\alpha\alpha'}$$
$$Q_{\mathbf{k},\alpha} Q_{\mathbf{k}'\alpha'} - Q_{\mathbf{k}'\alpha'} Q_{\mathbf{k}\alpha} = 0$$
$$P_{\mathbf{k},\alpha} P_{\mathbf{k}',\alpha'} - P_{\mathbf{k}',\alpha'} P_{\mathbf{k},\alpha} = 0.$$

But, clearly, we can now define linear combinations, of the P and Q operators, which are operator analogues of the Fourier coefficients $c_{\mathbf{k},\alpha}$ and $c^{*}_{\mathbf{k},\alpha}$:

$$a_{\mathbf{k},\alpha} = \frac{1}{\sqrt{(2\hbar|\mathbf{k}|c)}}\,(|\mathbf{k}|c\,Q_{\mathbf{k},\alpha} + iP_{\mathbf{k},\alpha})$$

$$a^{\dagger}_{\mathbf{k},\alpha} = \frac{1}{\sqrt{(2\hbar|\mathbf{k}|c)}}\,(|\mathbf{k}|c\,Q_{\mathbf{k},\alpha} - iP_{\mathbf{k},\alpha}).$$

The properties of the operators a, $a^{\dagger}$ and the product $N = a^{\dagger}a$ are such that they provide a natural description of photon states, including processes such as absorption and emission of radiation wherein the number of photons with given wavelength (or momentum) and polarization is increased or decreased. If we represent a general situation, where many different photon states are represented, by a row vector

$$n_{\mathbf{k}_1\,\alpha_1} \cdot \cdot \cdot \cdot \cdot \cdot \cdot \cdot \cdot n_{\mathbf{k}_i\,\alpha_i} \cdot \cdot \cdot \cdot \cdot \cdot \cdot \cdot >,$$

where $n_{\mathbf{k}_i\,\alpha_i}$ is the number of photons occupying a given state, i.e., the number of photons of momentum $h\mathbf{k}_i$ (or wavelength $2\pi/\mathbf{k}_i$) and polarization state α_i, then the effect of operating on this row vector by $a^{\dagger}$, a, and N, respectively, will be to increase by one, decrease by one, and leave unchanged the occupation number for a given state $\mathbf{k}_i,\alpha_i$.

Let us now return to our original Fourier expansion for the electromagnetic potential $\mathbf{A}(\mathbf{x},t)$. We can now replace the Fourier coefficients c and c^{*} by their operator analogues a and $a^{\dagger}$. The field quantity $A(x,t)$ is then itself an operator, sometimes called a field operator or a quantized field. We can easily see that the energy (H) and momentum (P) of this quantized radiation field are

$$H = \Sigma_k \Sigma_\alpha \, \hbar c |k| N_{k,\alpha}$$

and

$$P = \Sigma_k \Sigma_\alpha \hbar \, k \, N_{k,\alpha}.$$

That is, the energy of the system is equivalent to that of a system of $N = \Sigma_k \Sigma_\alpha N_{k,\alpha}$ oscillators with $\Sigma_\alpha N_{k,\alpha}$ of them having energy $h\,c|k| = h\nu$, just as required in the analyses of Planck and Einstein. Substituting the expressions for the energy and momentum of an individual photon into the relativistic energy equation,

$$E^2 = p^2 c^2 + m^2 c^4,$$

we see that the mass of the photon is identically zero, just as expected. Furthermore, the rotational transformation properties of the polarization vectors ϵ^α are formally analogous to those of the vector representing the angular momentum of particles with spin one. These results are often expressed by saying that "quantum mechanical excitations of the radiation field can be regarded as particles of mass zero and spin one." It is further held that this result reflects a general feature of quantum field theory—every (type of) particle is an excitation of some particular field and, conversely, with every field we can associate particles of definite mass and spin (cf. Sakurai 1967).

In discussing material particles, it is important to remember the difference between fermions and bosons, particles which are described, respectively, by Fermi-Dirac and Bose-Einstein statistics. For present purposes, the important difference between the two types of particles is in the number of particles which can occupy a given state. At most, one fermion can occupy a given quantum mechanical state; the occupation number for fermion states is either zero or one. Any non-negative integer number of bosons, on the other hand, can occupy a given state. Furthermore, there is a simple relation between the intrinsic spin of particles and their statistical behavior; particles with integer $(0,1,2 \ldots)$ spin values are bosons, while particles with half-integer $(1/2,3/2,5/2 \ldots)$ spin values are fermions.

158

Relativistic and Field Theory Considerations

It is instructive to notice how these points are related to the difficulties of the unquantized relativistic "wave-field" equations discussed in the preceding section. The difficulties for both the spinless particle Klein-Gordon equation and the spin 1/2 particle Dirac equation are ultimately traceable to the existence of negative energy solutions and their corresponding wave functions; this point can be generalized to include the factorizations of the Klein-Gordon equation corresponding to particles of any given spin, whether integer or half-integer, i.e., whether bosons or fermions. For Dirac particles, and fermions generally, we were able to evade the difficulty, temporarily, by postulating a sea of filled negative energy states, but even this temporary evasion is unavailable for bosons. No matter how many bosons occupy a negative energy state, there is always "room" for more. Dirac's postulate would be futile as an attempt to preclude *boson* transitions to negative energy states. But now notice that the formalism sketched above for the quantum theory of radiation is just that appropriate for bosons. The photon "particles" involved have spin one, and the properties of the operators a, a^t, and N and the state vector $| \ldots n \ldots >$ are such that any non-negative integer occupation number is allowed for any given state. Hence with no essential modifications, this same formalism can be applied to boson-particle "fields"—i.e., to the solutions of the corresponding wave equations.

For example, spinless π meson particles are described by the Klein-Gordon equation. Just as we quantized the vector potential $\mathbf{A}(\mathbf{x},t)$ characterizing the electromagnetic field by expanding it in a Fourier series and replacing the coefficients by operators satisfying certain commutation relations, so too we can quantize the Klein-Gordon "field" by expanding the wave function solution $\psi(\mathbf{x},t)$ of the Klein-Gordon equation in a plane wave series and replacing the coefficients by operators satisfying certain commutation relations. Then, again analogous to the radiation case, we can show that the energy and momentum of the Klein-Gordon "field" are equivalent to the energy and momentum of a collection of particles. From the rotational transformation properties of the wave function, ψ, we know that the particles are spinless. (This analysis applies for both charged and uncharged Klein-Gordon particles. Neutral π mesons, e.g., are described by a real-number-valued wave function; charged π mesons by a complex-valued wave function which is a linear combination of real-valued functions. For a π meson particle of positive charge, the negative-energy wave functions describe its negatively charged antiparticle, and conversely.)

We cannot apply the above procedure in straightforward fashion to the fields, i.e., the wave functions, describing fermions. If we adopt the above formalism we will have a theory which permits the occupation of a given state by any number of particles, a theory inapplicable to fermions. It is a remarkable fact, however, that a relatively simple modification of the formalism allows use of the same general approach for fermions as for bosons. That is, if we assume that the creation and annihilation operators a and $a^{\dagger}$ (the operator analogues of the Fourier coefficients) satisfy anticommutation relations rather than commutation relations, then everything proceeds as before. In particular, if

$$a_{\mathbf{p}}^{(r)} \, a^{\dagger}_{\mathbf{p}'}{}^{(r')} + a^{\dagger}_{\mathbf{p}'}{}^{(r')} \, a_{\mathbf{p}}^{(r)} = \delta_{rr'} \, \delta_{\mathbf{p}\mathbf{p}'}$$

and

$$a_{\mathbf{p}}^{(r)} \, a_{\mathbf{p}'}{}^{(r')} + a_{\mathbf{p}'}{}^{(r')} \, a_{\mathbf{p}}^{(r)} =$$

$$a^{\dagger}_{\mathbf{p}}{}^{(r)} \, a^{\dagger}_{\mathbf{p}'}{}^{(r')} + a^{\dagger}_{\mathbf{p}'}{}^{(r')} \, a^{\dagger}_{\mathbf{p}}{}^{(r)} = 0$$

for pairs of fermions of momenta $\mathbf{p}$ and $\mathbf{p}'$ and spin-energy indices $\mathbf{r}$ and $\mathbf{r}'$, then the resulting formalism permits only values 0 and 1 for state occupation numbers. Then the energy and momentum of the quantized field or field operator $\psi(\mathbf{x},t)$ written in terms of these anticommuting creation and annihilation operators are seen to be equivalent to the energy and momentum of a system of particles.

Even from this brief sketch of the quantum theory of fields we can see the plausibility of the equivalence claim in question. Whatever physical systems we are interested in, whether they are radiation fields or collections of material particles, bosons or fermions, the most generally correct description of these systems proceeds in strikingly similar fashion. We always begin with the wave equation describing the particular system of interest, expand the solution of this equation in a plane wave series, replace the expansion coefficients by operators interpretable as creation and annihilation operators, and find that the energy and momentum of the quantized "wave-field," as calculated by standard prescriptions which are operator analogues of classical functions, are equivalent to those of collections of particles.

This common field-theoretic procedure is designed so as to naturally handle processes in which particles, whether material particles or photons, are created or destroyed. Hence it corrects the deficiencies of both the nonrelativistic Schrödinger theory and the unquantized relativistic theories. Given a formalism of this scope, power, and elegance, it is not surprising to find many, if not most, authors attracted to a point of view in which the physical interpretation of the formalism is the same no matter what kind of physical system the formalism is applied to. The claim that every particle is an excitation of (a mode of oscillation of) some field and, conversely, that with every field there is to be associated some characteristic type of particle, is a claim which is attractive precisely because of its simple, general, systematic character. However attractive the claim may be, it is not one which can be taken very seriously. As a remark about formalisms and their manipulation it is unassailable; as a remark about physical processes it is very misleading in that it leaves most of the story untold—it ignores differences which are far more striking and important than the cited similarities.

Heuristically, it is convenient to begin our criticism by enumerating some of the more important differences between photons and fermions, putting aside for the moment (material) boson particles. The most important single difference between the quantized radiation field and quantized fermion "fields," e.g., Dirac fields, lies in the nature of the field quantities which are quantized. The radiation field characterized by the vector potential $\mathbf{A}(\mathbf{x},t)$ is a physical system which, even in the absence of particles acting as sources and/or sinks of radiation, possesses actual energy and momentum. Unless this fact is accepted, a whole host of classical phenomena become inexplicable. This is why we are able to express the radiation-field creation and annihilation operators and through them the field quantity $\mathbf{A}(\mathbf{x},t)$ itself as linear combinations of the canonical coordinates and momenta, P and Q, satisfying certain commutation relations. Quantum mechanically, physically measurable quantities are those which figure in such commutation relations. Hence quantum mechanically, just as classically, the field quantities $\mathbf{E}$ and $\mathbf{B}$, related to their surrogate $\mathbf{A}$ via $\mathbf{B} = \Delta \times \mathbf{A}$ and $\mathbf{E} = -\frac{1}{c}\partial\mathbf{A}/\partial t$, are physically measurable quantities.

The situation is quite different for the fermion wave function $\psi(\mathbf{x},t)$. Here we are not able to express the creation and annihilation operators as linear combinations of canonical variables satisfying commutation relations. This inability is no mere happenstance. For if we could so express those operators, the fermion wave function $\psi(\mathbf{x},t)$ would be a measurable field quantity. But we know that in the unquantized fermion "field" theory, it is the current or flux of particles which is measurable and, of course, we have constructed our quantized field theory so that it is the corresponding current operator rather than the field operator $\psi(\mathbf{x},t)$ which satisfies commutation relations and, accordingly, is the measurable quantity.

The above remarks indicate that the wave-field aspects of electromagnetic radiation are to be regarded as actual physical properties while the "field" function for fermions is to be regarded, just as in the nonrelativistic Schrödinger theory or the unquantized Dirac theory, as an abstract, mathematical item—one related to the probability distribution of particles when their numbers are conserved and, more generally, to the probability of creation, annihilation, and preservation of particles. Just the opposite attitude is indicated for the "particle" properties of radiation as compared to the particle properties of fermions. Recall the remarks made at the end of our discussion of classical radiation theory. There it was pointed out that the analogy between modes of oscillation of radiation fields and material particles, particles described by mechanical laws, is a very limited and essentially formal analogy. In particular, it is the wave or field disturbance, not the fictitious "particles," which travel through space, and this point still applies in the quantum theory of radiation.

When we quantized the electromagnetic field by replacing the classical functions $\mathbf{A}$, $\mathbf{E}$, and $\mathbf{B}$ by corresponding operators satisfying certain commutation relations, we in effect introduced an indeterministic theory for the description of the field. Just as the quantization of classical mechanics leads to indeterminism, to uncertainty relations between certain variables, to discrete energy spectra and discrete energy transitions for certain systems of particles, so quantization of electromagnetics leads to uncertainty relations between field strengths and photon occupation number, as well as between components of $\mathbf{E}$ and of $\mathbf{B}$, and to discrete energy and momentum transfer between radiation fields and systems of particles. In particular, if the photon occupation number for a particular mode of oscillation is well known then the uncertainty in field strength will be very high; conversely, if the occupation number is highly uncertain, the field strength can be known to high accuracy.

This last point bears on the relation between classical and quantum radiation theory. Just as classical mechanics can be shown to be a special, limiting case of quantum mechanics, so too classical radiation theory can be shown to be a special, limiting case of quantum radiation theory—the average or expectation values of the operator analogues of the classical field variables can be shown to satisfy Maxwell's equations; and when the photon occupation numbers for particular modes of oscillation are so large that they may as well be treated as continuous (real numbered) rather than as discrete (integer valued) variables, then the classical description is indistinguishable from the quantum description. Notice the difference in the "direction" of the classical limit in the two cases of radiation fields and material particles. For material particles the classical description is correct for a single particle; for radiation, the classical description is correct for an essentially infinite number of "particles," i.e., photons, per unit volume.

This difference is most revealing and suggestive. It reinforces our earlier suspicion that the photonic "particles" of radiation theory are not particles at all. They are not even approximately localized or localizable. A given photon can exist simultaneously at all points in an extended volume of space, a volume whose time rate of change is proportional to the speed of light. Of course, the disturbance or mode of oscillation associated with a photon travels through space, but it is the wave-*front* which travels; in general, the photon, i.e., the oscillation, continues to exist at all points between the source and the (component of the) wave-front. One source of confusion about photons arises from the obvious fact that they are emitted at particular locations and absorbed at other particular locations; since this is so, it is natural but mistaken to suppose that they travel as particles, as localized carriers of momentum and energy, from source to sink, from emitter to absorber. This not unnatural supposition is mistaken because, as noted, they exist at all points between source and wave-front; furthermore, when photons are absorbed, the damping effect of the absorber on the field oscillation is transmitted back through the field at the speed of light, not instantaneously. These elaborations of earlier points amount to no more than a reminder that a photon, unlike a material particle, occupies many spatial points simultaneously—theoretically it could simultaneously occupy all space. The classical limit theorem of quantum radiation theory reminds us that, again in contrast with material particles, many photons can, simultaneously, literally occupy the same point in space.

As noted, the indeterministic quantum description of material particles leads to the prediction of discrete energy levels and discrete energy transitions or transfers. Similarly, the indeterministic quantum description of radiation fields leads to the prediction of discrete interactions, i.e., energy transfers, between radiation fields and material particles. Thus it correctly describes, predicts, and explains such previously puzzling phenomena as the blackbody radiation spectrum, the photoelectric effect, the Compton effect, pair creation and pair annihilation, among many others. The analogy between the proper interpretation of the quantum mechanics of material particles and the proper interpretation of the quantum theory of radiation is quite extended. Roughly speaking, the two interpretations are the same except that the roles of the terms 'particle' and 'wave-field' are reversed. In the material particle theory, the particle aspects of physical systems retain their essential classical role, the possession of simultaneous position and momentum, but the blissful classical presumption of determinism must be abandoned. Particles travel in continuous paths, but it is only the statistical distribution of these paths and not the individual trajectories that are determined by mechanical laws. The "wave-field" aspects of material particles have nothing to do with the actual spatiotemporal oscillations of any physically measurable quantity, and are explained as abstract items related to probability distributions.

In the radiation theory, on the other hand, it is the wave-fields which are physically measurable and which retain the classical aspect of being continuously distributed over large spatial regions at a given time. The particlelike properties of the radiation field have little in common with those of material particles, reflecting nothing more than discrete interaction properties of the field, predicted by the indeterministic description implicit in the formalism of quantum radiation theory.

Note carefully this relation between the discrete interactions and the indeterministic description. The formalism of quantum radiation theory entails, besides those uncertainty relations already cited, an uncertainty relation between the photon occupation number of a given state—i.e., the intensity of a given mode of oscillation—and the phase of this oscillation—i.e., the spatiotemporal coordinates of the (plane) wave-front. Hence if the total momentum of a wave component is well known, the position of the wave-front is poorly known and vice versa. Accordingly, when we describe particle-field interactions as occurring at a definite spatiotemporal location the results necessarily show up as the transfer of the energy and momentum corresponding to one or more photons. Hence our indeterministic formalism is sufficient to guarantee the discrete transitions required by a variety of observational evidence; conversely, the discrete transitions are a necessary consequence of the indeterministic formalism. (Finally, notice that, if the cited phase-intensity uncertainty relation were not valid, we could measure the simultaneous position and momentum of material particles; light beams of exactly known intensity and phase could be used to fix the particle exactly at two different times, and the momentum transferred to the particle at these times would be inferrible and correctable for.) To summarize, the "particle" properties of light beams or other electromagnetic wave-fields are reflections of discrete interaction properties, and these in turn are reflections of the basically indeterministic character of physical processes involving radiation fields.

This last point is a crucial one and requires amplification. Recall that even classically there is a reciprocal uncertainty between the wavelength and the wave-front location of a plane wave. This relation holds completely apart from any quantum mechanical considerations, but this classical uncertainty relation does not preclude simultaneous measurement of particle position and momentum or, equivalently, does not preclude continuously controllable interaction between the wave-field and the particle. For, classically, the wave could carry any amount of momentum, an amount independently ascertainable, and interaction of wave and particle could be minimized by using waves of very low momentum and energy content; then accurate particle position measurements could be made without disturbing the particle momentum. Of course, in obtaining an accurate specification of the wave-front location we have lost all accuracy in our wavelength specification, but in the classical picture this fact is harmless. So long as waves of any given length can take on any energy and momentum values, we can arrange for all wavelength components of our incident wave to carry as much or as little momentum as we please. Hence we can know that the whole wave carries negligible momentum without knowing anything at all about its wavelength.

It is only when we introduce the quantum condition, $p = h/\lambda$, relating wavelength and momentum, that our wavelength uncertainty becomes a momentum uncertainty and precludes an accurate simultaneous measurement of conjugate variables. So far it may seem that these points about uncertainty relations, i.e., about the indeterministic description, have nothing to do with the matter of discrete interactions, but this appearance is misleading. To see the connection it is helpful to invoke the formal analogy between radiation fields and harmonic oscillators. Classically, each of these oscillators can take on any momentum value. Hence when the field comprised by the total collection of such oscillators interacts with a system of particles, the momentum (and energy) gained or lost by any or all of these oscillators can be arranged to be as large or small as we please. But what happens if the momentum and energy of each oscillator is fixed forever, i.e., if wavelength and momentum are uniquely related? Obviously, any interaction between the field and another system that shows up as a change in field energy and momentum will necessarily be described as the creation or annihilation of one or more oscillators of certain momenta and energies, or wavelengths and frequencies.

So an extremely simple analysis establishes a point not sufficiently emphasized in the literature: the discrete interactions characteristic of quantum radiation theory and required for prediction and explanation of a variety of experimental results are an immediate consequence of the indeterministic description built into the theory.

Before leaving the present topic, one final comment is in order. For the most part we have analyzed radiation fields and collections of material particles separately in this study, and our separate analyses have revealed that discrete interactions are characteristic of both of these types of physical systems. Naturally, this is no coincidence. The correct description of these two types of systems must mesh, must match up. If material particles undergo transitions between discrete energy states and thereby emit or absorb radiation, the radiation fields involved must be capable of corresponding discrete transitions in order for us to have a consistent description of the total processes.

There are other important differences between quantized radiation fields and quantized fermion fields. We were able to derive the zero mass characteristic of photons from the quantized field theory results for photon energy and momentum, together with the relativistic momentum-energy relation. We are able to use this procedure for photons precisely because the quantities v and λ occurring in the expressions for energy and momentum correspond to the frequency and wavelength of actual physical waves and are related by $\lambda v = c$, where c is the speed of these electromagnetic waves.

No such procedure is available for fermions. The energy and momentum of fermion "fields" are simply the sums of the energies and momenta of the particles involved, but there is no equation relating the energy and frequency of a fermion wave. The frequency which occurs in the relation $E = hv$ is the frequency of electromagnetic radiation emitted by systems of particles making transitions to lower energy states—it does not correspond to any "frequency" of Schrödinger or Dirac waves; similarly, there is no frequency for the "wavelength," $\lambda = h/p$, of Schrödinger or Dirac waves to be related to.

Putting the matter another way, there is no unique relation between energy and momentum for material particles as there is for photons; the energy of a particle of a given momentum depends on the mass of the particle; hence there is no hope of calculating the mass from any unique relation between energy and momentum. We might have anticipated this result simply by remembering that there are many different fermions or particles with half-integer spins found in nature—electrons, protons, neutrons, muons, neutrinos, etc.—and they vary greatly as to their masses. If there is to be a one-one correspondence between particles and fields, we must distinguish between the Dirac field for electrons, the Dirac field for protons, etc., even though the formalisms for these fields are indistinguishable. There seems to be no good reason for this proliferation of fields, beyond the desire to save the sweeping thesis under attack.

Furthermore, we must be skeptical of the claim that particles are modes of oscillation or "excitations" of the particular fields they correspond to. This way of expressing the claim makes it seem as though the particle acquired its energy from the field and could, upon the particle's destruction, transfer the energy back to the field. There is some gross confusion involved here. Photons are indeed excitations of their corresponding field—but then we have seen that photons are not particles at all, except in a very fanciful sense. Fermions, e.g., electrons, are not excitations of the Dirac field except in a formal sense; the Dirac field carries no energy or momentum apart from the energy and momentum of the fermion particles involved. When fermion pairs are created, the energy required for this process comes from the electromagnetic field, not from the Dirac field! Conversely, when fermion pairs annihilate each other, the resulting energy appears as an increase in the energy of the electromagnetic field, not the Dirac "field."

These remarks indicate that even the briefest reflection on the physics of the situation will bring to mind important differences in the roles played by the mass of material particles and by the (non-existent) mass of photons. Further reflection yields an important difference between the spin of material particles and the "spin" of photons. The spin of material particles is an actual physical effect, a *rotation* about some axis of the particle. This spin is over and above the *orbital* or *revolutionary* angular momentum which the particle possesses with respect to some axis outside the particle. The claim that photons possess one unit of "spin" is quite misleading, since the angular momentum which takes on integer values for photons is the total angular momentum, i.e., orbital plus spin angular momentum. For the case of a plane wave expansion of electromagnetic field oscillations, this total angular momentum is entirely composed of orbital angular momentum; there is no angular momentum corresponding to the intrinsic spin of material particles. We could expand the field oscillations in spherical harmonics or even, heaven forbid, in cylindrical functions. Here, again, it is the total angular momentum which has integer values, and no clear breakdown into orbital angular momentum and spin angular momentum is possible. In no case can we identify a photon spin which corresponds to the intrinsic spin of material particles. The claim that "photons have spin one" must be interpreted to mean that the rotational transformation properties of the matrix representing the *total* angular momentum of the photon or plane wave mode of field oscillation are formally identical to those of the matrix representing material particles of spin one.

Some of the points we have made regarding fermions and photons, and their corresponding fields, can be restated as points about the usage of the term 'field.' In the most attenuated sense, a field is simply a continuous function of one or more spatial variables—a function defined at every point of some region of space. In this sense the Dirac fermion field is just as much a field as the electromagnetic field, but in this attenuated sense the term 'field' tells us nothing whatever about what kind of physical item the field function represents. There is a second, less attenuated sense of 'field' in which a field function at a point represents the magnitude and direction of a particular kind of force which bodies of a particular sort experience when located at this point. Clearly, fermion fields are not fields in this second sense. Neither the Dirac wave function nor any bilinear form of it represents such a force. Both gravitational fields and electromagnetic fields are fields in this second sense, but electromagnetic fields are also fields in a third and still stronger sense of 'field.' Electromagnetic fields are, but fermion fields and even gravitational fields are not, radiation fields.

Such radiation fields possess actual momentum and energy over and above the momentum and energy of material particles. Systems of material particles act as sources of this energy and radiate it to the field, which can store energy and momentum for indefinite periods or transmit it to other systems of material particles. This energy-storage feature of radiation fields is essential for an explanation of various phenomena in classical electrodynamics. One way of bringing out the difference between electromagnetic and gravitational fields is to note the precise sense in which the former are theoretically indispensable while the latter are not. Classically, gravitational fields are invoked to explain certain motions of particles; knowledge of the field—of the spatial distribution of forces—is sufficient for prediction of the motion of particles through space. But the field at any time is itself determined by the spatial configuration of masses at that time. Hence the field is a dispensable intermediate variable; we use the configuration of masses to find the field and then use the field to find future configurations of particles, but we could short-circuit the process and go directly from one configuration of masses to future configurations.

The situation is quite different in classical electromagnetics. There knowledge of the configuration of (charged) particles is not sufficient to determine the field. We must have independent knowledge of the field strengths, which can be quite different from those set up by the configuration of charged particles. In effect we must know how much energy is stored in the field, how much (net) energy was previously radiated to the field by systems of particles. In short, we must know the entire history of the total system— particles plus fields. The arguments of this chapter amount to a reminder that the classical distinction between two kinds of physical systems, two quite dissimilar items which carry energy and momentum, particles and fields, remains a valid distinction even after the required quantum mechanical corrections to classical physics are included. Quantum mechanically, just as much as classically, we have good reason to believe that electromagnetic fields are fields in the third and full-blown physical sense of 'field' while no other fields are.[3] Conversely, fermions are genuine classical particles while photons are particles only in a fanciful and misleading sense.[4]

To complete our criticism we need only cover the case of material boson particles, e.g., π mesons. Superficially, such boson particles bear strong resemblance to both of the disparate items discussed above, fermions and photons, and their corresponding fields. Material boson particles resemble photons in their spin values and, accordingly, in their statistical behavior. Many bosons can occupy the same quantum mechanical state. Yet such boson particles also are clearly just as much material particles as are fermions since they have nonzero mass values. Once again, a little reflection will convince us that bosons are genuine material particles like fermions, that the similarities between material particles and photons are formal in character, and that these similarities are misleading if they inspire us to assume more substantial resemblances.

To begin with, we can no more calculate the mass of material boson particles from the quantized field energy and momentum expressions than we can for fermions. Once again, there is no unique relation between particle energy and momentum or, equivalently, there is no measurable radiation field associated with systems of material boson particles, no actual physical waves with parameters ν and λ connecting E and p. Just as with fermions, there are many, say, mesons of different mass for a given integer spin value. Is there then a different "field" associated with each spin-one meson? Proliferation of distinct fields seems to add nothing to the conversation beyond what is already expressed by noting the differences among the particles in question; since none of the fields is a measurable radiation field anyway, it is idle to speculate on whether one or many undetectable fields are present.

Another, and perhaps crucial, point is that the classical limit of the Klein-Gordon equation or any of its integer spin factorizations, i.e., the equations describing bosons, is the nonrelativistic Schrödinger particle equation and, via this latter equation, ultimately, the classical mechanics of a single particle; this is exactly as for the fermion equations represented by the Dirac equation or other half-integer spin factorizations of the Klein-Gordon equation. On the other hand, the classical limit of the quantized field equations for photons is Maxwell's equations, i.e., equations for a measurable radiation field.

So it is clear enough that bosons, other than photons, are genuine material particles, just as much as fermions. In view of this it may seem puzzling that the spin values, and hence the statistical behavior, of material boson particles are those of photons rather than fermions. We have already noted the formal character of the analogy between photon spin and material particle spin; actually, the photon "spin" is a classical effect, referring to the orbital, not rotational, angular momentum of (a section of) a plane wave with respect to some axis. Since this is so, and since we know the definite connection between spin and statistics, we should suspect that the similarity between the statistical behavior of photons and that of other bosons is also a formal similarity, a similarity masking important differences. This is easily seen to be the case. What photons have in common with, say, π mesons or even-(mass)-numbered nuclei is that any number of either type of particle can occupy the same quantum mechanical state; but the term 'state' refers to a quite different array of properties when applied to photons rather than mesons or nuclei. In the case of a photon, the physical quantities defining a state are the momentum and polarization vectors; but these vectors characterize a plane wave extended throughout (potentially) all space; any given photon occupies many spatial locations simultaneously and, conversely, many photons can occupy the same location. Thus it is trivially obvious that photons are described by Bose-Einstein statistics; more precisely, the wave function describing an assembly of photons is symmetric, or un-changed whenever the vectors characterizing any pair of photons in the same "state" are interchanged.

This situation is in sharp contrast to state descriptions of mate-rial particles, whether fermions or bosons. Assemblies of "states" of material particles are typically characterized by the spatial and spin coordinates of each of the particles of the assembly. The differ-ences between fermions and bosons lies in the different behavior of the wave function describing such an assembly when the spatial and spin coordinates of any pair of particles are interchanged. If the particles are fermions, the new wave function will be dis-tinguishable from the original because of the transformation prop-erties of the matrices representing half-integer spins. So long as the fermions are not interacting with each other, we can distinguish among them and say they are in distinct states. If the particles are bosons, on the other hand, interchange of the coordinates of any pair will have no effect on the wave function. Hence there will be no way, experimentally, to distinguish between the pair, since the wave function contains all available information about the assembly of particles. (Again, the symmetric character of the wave function is related to the transformation properties of the integer spin matrices.) So we say that particles described by such symmetric wave functions are all in the same state. But to say we cannot dis-tinguish between one massive particle at location r_1 and another at r_2 is quite different from saying we cannot distinguish between two photons both occupying all points of a spatially extended region. In the former situation our description presupposes that one par-ticle is at one location, the other particle at the other location, but we do not, and cannot, know which is which. In the latter case our description presupposes that both "particles" are at all locations in the region.

A more complete discussion of bosons than we can undertake here is complicated by the fact that there are measurable force-fields, i.e., 'fields' in the second sense specified above, associated with at least one boson particle, the π meson. The π meson is vaguely "associated" with the strong binding forces among nucleons, forces holding nuclei together in spite of the strongly repulsive electro-static forces among protons. Such fields are not radiation fields; they are extremely short-range fields, of the order of 10^{-13}cm. Hence any extranuclear "radiation" associated with this field must show up as either electromagnetic radiation or simply particle trajectories, e.g., mesons. Indeed, the only theoretical estimate of pion mass is based on the form and range of this force-field and not on the energy and momentum of the quantized field. A thorough investigation of the precise relationship between pions and intranuclear force fields would be most worthwhile but beyond the scope of the present study. We must confine ourselves to the weak claim that there is nothing presently known about mesons and intranuclear forces which precludes or even counts against the distinctions made earlier. On this note, we conclude our discussion of quantum field theory. This final chapter has been intended as no more than a sketch of relativistic and field theory considerations, which deserve a treatment as fully extended as the earlier treatment given the nonrelativistic Schrödinger theory.

Difficulties of the Present Theory

In concluding, I will mention the difficulties plaguing the quantum field theory discussed in the previous section, the most generally correct theory of physical processes. This recital of difficulties is included for two reasons. First, there is the consideration of completeness. These difficulties are, after all, well-known aspects of the same theory whose other aspects are of direct interest to philosophers. It would be gratuitously risky to discuss any subject matter while deliberately ignoring certain aspects of the subject. Second, the difficulties in question remind us, if we need yet another reminder, that no physical theory, no matter how general, powerful, or successful, is likely to retain a position of superiority forever. No theory, if the future is anything like the past, can be expected to tell us the truth, the whole truth, and nothing but the truth about the world. Just as the present theory replaced previous, relatively inferior theories, so the present theory will eventually be superceded by a newer and better theory. The difficulties with the present theory point to the urgent need for such a new theory.

We have already passed over in silence one of the difficulties of quantum field theory: the energy of a quantized radiation field can be shown to be the sum of terms of the form

$$E = h\nu.$$

Strictly speaking, this is misleading. A straightforward calculation yields

$$E_\nu = h\nu \, (N_\nu + 1/2)$$

for each term, i.e., for the energy of each radiation oscillator. We tacitly assumed that energy measurements are relative, not absolute, and that the vacuum, or field with no occupied states, has energy $(1/2)h\nu$ for each state but that this infinite amount of energy is undetectable. This is a painless way out of our difficulty and works for fields with no sources or sinks present. When charged particles are present, however, it is not so easy to use this trick in a consistent way.

This latter fact is related to a difficulty which arises for charged particles even in the absence of externally applied fields. Charged particles are never free of fields: the very fact that they are charged means that they act as sources of electromagnetic fields and then, again because they are charged, interact with these self-created fields. The energy of this self-induced interaction must be included as a correction to the Hamiltonian used in calculating various quantities of interest in the interaction of the particles with other systems—e.g., transition probabilities for emission, absorption, scattering, etc. But then a troublesome result occurs: this correction term involves summation or integration over all possible states of the self-created radiation field, and this summation or integral diverges. To avoid this unpleasant result, theoreticians arbitrarily cut off the summation or integration at some high energy-value. The use of such a cut-off energy as an upper integration or summation limit is equivalent to applying a correction factor to the momentum or, equivalently, the mass of the particle.

Let us return to isolated radiation fields for a moment. A correction term for the effective "mass" of such fields is to be expected since, in general, part of the field energy can be used to create charged particle pairs. Formally, this mass correction term is equivalent to a correction factor for the electric charge, since all the radiation field interactions are proportional to this charge. Once again, an energy cut-off is used to obtain the requisite correction factor.

There are three different reasons why a theory dependent on such cut-off methods must be regarded as fundamentally unsatisfactory.

First, physical processes can occur at all energies, no matter how high. Hence the theory cannot possibly account for all processes. Second, the higher the cut-off, the larger are the correction factors for particle charge and mass; and extremely high cut-offs are required for non-negligible corrections. In order for a significant portion of the observed rest energy or rest mass of the particle to be attributable to the postulated effects, the energies involved in familiar processes must be tremendously higher than we have ever had any reason to believe. This renders the entire theory quite suspect. Third, and most disconcerting of all, the form of the required correction factors is such that the resulting Hamiltonian-wave function formalism cannot be guaranteed to be consistent. The sum of the probabilities for all possible physical processes in a given situation will not, in general, be guaranteed to be equal to one. This embarrassment can only be avoided by giving up features of theoretical physics equally as cherished as a consistent probabilistic interpretation of quantum mechanics—the requirement that all basic physical laws be Lorentz invariant, i.e., relativistically correct; or that all interactions be localized, whether particle-particle, particle-field, or field-field interactions, i.e., no instantaneous action at a distance.

It would be worthwhile to trace out in detail the origins, interconnections, and philosophical implications of the difficulties of the currently prevailing theory. Unfortunately, such a project is far beyond the scope of the present study. The story of the difficulties of contemporary quantum field theory, while fascinating, is a story for another day.

Notes

Chapter 1

1. Solutions of the Schrödinger equations for particular situations figure importantly in the discussions of later chapters. See especially the classical limit theorem (chap. 2) involving the time-dependent Schrödinger equation.

2. Schrödinger's later views are not essentially different from his original ones. See Schrödinger (1952).

3. We shall see that this claim is misleading, that the implied analogy between photons and their associated wave function on the one hand, and material particles and their corresponding function on the other is far from complete. Furthermore, the above argument equivocates on the type of wave function under discussion—the claims of Bohr and Heisenberg apply only to the elementary harmonic solutions of the wave equation and not to superpositions of these.

4. These statements must be interpreted cautiously. Even if we always measure the same eigenstate there will be some dispersion among the numerical values obtained. But in general we are even worse off than that—we do not even know which eigenstate we will be measuring. Thus we must distinguish two senses of "dispersion," one referring to the uncertainty as to which eigenstate will be obtained, the other to the uncertainty in the exact value of a given eigenstate.

Chapter 2

1. See Bunge (1967), a collection of technical and critical essays on quantum mechanics, with emphasis on the Copenhagen interpretation, and containing the papers of Bunge and Popper which are discussed in this chapter.

2. Unfortunately Bunge, Popper, et al. never make it as clear as we would like *exactly* what view they are attributing to the Copenhagen interpretation. Accordingly, I cannot hope to make clear what is not clear. It is sufficient for my purposes to show that whatever the view Bunge and Popper think they are attacking, it is *not* the Copenhagen interpretation.

3. It has been shown that every syntactically consistent formalism has an interpretation in the realm of natural numbers. See Kleene (1952), p. 398, for statement and proof of this meta-mathematical theorem.

4. Of course, "minimal semantic" connections might fall short of sameness of meaning. But we cannot have it both ways: given any pair of theories, as opposed to uninterpreted schema, one cannot legitimately be derived from the other unless the common terms mean the same in both. Derivations (i.e., deductions) involving equivocations are fallacious. If one theory is a limiting case of the other in some other, weaker sense, then variation in meaning is possible. Conversely, if such meaning-variance is independently established, then it follows, with apodeictic certainty, that a weaker sense of "limiting case" is the best we can hope for. My strategy, obviously, is to head off the latter possibility by independently establishing the meaning-invariance of the key common terms. So far I have only promised, not delivered, the actual arguments, but they begin within the next few pages.

5. Notice that this assumption breaks down at classical turning points— where the direction of motion of particles changes discontinuously. Of course, these discontinuous changes only occur in processes involving theoretical idealizations—perfectly rigid bodies; no actual body even approaches perfect rigidity.

6. The above discussion is strictly correct only for the nonrelativistic case. The formulation of the argument for the general case where the mass is velocity-dependent is more complicated but the essential point remains unchanged. (If the particle is charged the situation is still more complicated but, again, the same conclusions can be reached.)

Chapter 3

1. It is quite clear that Nagel *does* claim that every theoretical axiom and theorem is a logically necessary condition for correct application of its occurring terms; indeed, it is the fact that "theoretical" terms are "implicitly defined" in this way that serves to distinguish them from "observable" terms. Cf. Nagel (1961), p. 87.

2. These classical mechanical laws do not transform properly, according to relativistic requirements, if they are interpreted in straightforward fashion. To transform correctly they must be doctored up. See the discussion at the end of chapter 2.

3. We might wish to accept the claim that the second equation of each set represents an explicit definition of the term 'force'; that this term is being used differently in the three theories. But if the definitions are indeed explicit ones, the term can be eliminated and the three laws in each set restated without using the term. For example, read "absence of other bodies" for "absence of forces" in the first law of each set. We need not worry about the meaning of explicitly defined terms since these are eliminable. Of course, this point tells us nothing whatever about the meaning of the (allegedly) implicitly defined, uneliminable terms.

4. For example, by optical means.

5. Such views have been held explicitly by Feyerabend (1962) and Kuhn (1962) and, at least at times, seemingly by Hanson (1958) and Ryle (1956).

6. These criticisms, among others, are presented in more detail by Achinstein (1968), pp. 91–98, among others.

Notes

7. The theoretical/observable distinction is badly confused because not all theoretical terms are terms for unobservables and because it obscures other, more important distinctions—between objects and properties or relations; between objects and substances; between abstract and concrete terms; between causal and logical relations; etc. The problem of theoretical terms is largely bogus because, typically, terms are not introduced and used as uninterpreted terms in science. The "solution" of *partial interpretation* is unacceptable because no satisfactory criteria for coordinating definitions have ever been formulated. Those criteria offered to date trivialize theory construction and fail to distinguish empirical science from armchair speculation, metaphysical or other. See, e.g., Achinstein (1968), pp. 67–91, 157–60, 179–201.

8. See chapter 2 above for a discussion of the distinction between *syntactical* and *semantical* aspects of meaning or usage. Of course, when the theory-proper is interpreted in terms of a (familiar) model, the items corresponding to the theoretical terms will have properties other than the syntactically imposed ones, but these other properties, according to Nagel, have nothing to do with the implicitly defined *meaning* of these terms. Cf. Nagel (1961), p. 300.

9. See Kleene (1952), p. 398.

10. We can play Nagel's "implicit definition" game with any set of sentences. For example, begin with any set of obviously *contingent* sentences; replace the subject matter terms with uninterpreted letters, leaving only the formal—logical plus mathematical—terms interpreted. If we now say this formal "calculus" imposes logically necessary conditions for correct application of the uninterpreted-letter terms we have transformed a set of contingent, empirical statements into a set of necessary, a priori statements. Clearly, we have gone off the rails somewhere!

11. That is, "we" cannot escape this conclusion so long as we try to defend Nagel's original criterion or its generic equivalent. But this is only to say that such a criterion represents an absurdly strict necessary condition for meaning (correct usage) and hence an absurdly liberal sufficient condition for meaning-change. If we hold a more natural and sensible view on the meaning of scientific terms—that typically some of the sentences, equations, etc., in which they occur represent important criteria for correct usage but not necessary conditions—then we have no difficulty in escaping the cited unpleasant consequences.

12. See Nagel (1961), pp. 277–93, 305–12. Nagel is here following Margenau very closely. Cf. Margenau (1950).

13. In Nagel (1961), pp. 281–82.

14. It is ironic that part of the inspiration for a deterministic construal of quantum mechanics can be traced to a few remarks of Heisenberg who, in general, was not at all opposed to an indeterministic construal. See Heisenberg (1930), pp. 62–65.

15. Of course, transitions between certain pairs of states are ruled out on general grounds—conservation laws, symmetry, etc.—whether the spectrum of eigenstates is discrete or continuous.

Chapter 4

1. I have discussed the classical conception of a particle at some length in chapters 1 and 2. The concept of a field will be discussed in chapter 7.

NOTES

2. The reader is urged to remember that we are dealing with that very special situation where two competing hypotheses are both consistent with the same body of (allegedly supporting) evidence, that is, according to the usual inductive criteria of *quantity, variety,* and *precision* of evidence. It is only when these usual criteria are indecisive that we need or can invoke *simplicity, initial plausibility,* etc.

3. It is important that I interject some disclaimers here. First, there may well be situations where the evidence does not support one theory more than its competitors; I am not claiming that there is always some one theory which is supported, to the exclusion of its competitors, by the observed evidence. I am only claiming that such unique support is possible and not even rare, although only a detailed historical examination would reveal which situation is "typical." Second, the conceptual point I am making, about the connection between simplicity and empirical support or confirmation, holds independently of any metaphysical presuppositions about the existence and uniqueness of true theories describing physical processes. It might turn out, for some unforeseen reason, that no general theory about physical processes could possibly be exactly correct. But this would not preclude the possibility of comparing the relative support or confirmation of competing theories since, strictly speaking, confirmation is a relation between statements—between the statements of a theory and the statements expressing the observed evidence. Even if no theory were (exactly) true, we could still compare two "approximate" theories—not with the true theory, since by hypothesis there is no such animal, but with each other. What we would be judging in each case, in evaluating the relative confirmation of each theory, is whether the evidence statements are the sort which count as confirmation of the expectations we would have if we believed the theory to be true. Whether this theory or indeed any theory is actually true is a separate question. The point I am making is an epistemological point and its validity does not depend on the truth of any metaphysical presuppositions. Of course, if no one in fact believed that any theory was or could be true, then the conceptual point I am making would be counterfactual but nonetheless correct.

4. When we come to apply the criterion of simplicity to concrete examples, we will see that these "distinctions" are quite misleading. The superiority of descriptively simpler hypotheses ultimately has nothing to do with formal or mathematical aspects of the competing hypotheses. Further, the similarity between descriptive and ontological simplicity is far more important than the differences: in both cases, the simplest hypothesis is to be preferred to its competitors because it alone is consistent with the data, without benefit of additional, unsupported hypotheses. The crux of the argument in both the curve-fitting and perception cases, discussed below, turns on exactly this point.

5. The greater complexity of the neon-world hypothesis lies, of course, not in the greater number or variety of items of furniture postulated but in the postulation of creation and annihilation *events* and, consequently, in the added complexity of natural laws. For example, none of our familiar conservation laws could hold, else creation or annihilation events at one spatiotemporal location could be detected at other locations because of their indirect effects.

6. Bohm or other hidden variable theorists might disclaim the argument I have attributed to them above, saying that although they equate "E happens without cause" with "E happens for no reason," they do not take the latter to be equivalent to "there is no reason to believe E happens," etc. But then

what have they added to the conversation by calling their theory 'rational' and the standard indeterministic theory 'irrational'? If they are not making an epistemic point, if they are only switching the labels 'causeless' and 'irrational' without building any epistemological content into the latter term, then what they are doing needs, and deserves, no reaction at all.

7. Amusingly, almost everyone instinctively claims that indistinguishability of *causes* for different effects is a matter of temporary ignorance, but hardly anyone ever points out that an indeterminist can play the same game with *effects*. One can always say that if the same cause leads to indistinguishable effects on two different occasions, the indistinguishability is a "mere" matter of temporary ignorance. I am indebted to my colleague, Professor David Clarke, for pointing this out to me.

8. See von Neumann (1955), pp. 295–328, for the details. The proof turns on the notions "homogeneous" and "dispersion free." These adjectives describe functions representing physical properties of ensembles or collections of similar systems. For a given physical quantity, q, a function of q is "dispersion free" if and only if the expectation value of q^2 is equal to the square of the expectation value of q, where these expectation values are related to the probability distribution of q-values. A function of q is "homogeneous" if and only if a given expectation value of q is *not* representable as a linear combination of independent expectation values of q—independent of each other and of the given expectation value. Briefly, what von Neumann shows is that no homogeneous function—and hence no physical property of a corresponding ensemble—is dispersion free. For if a homogeneous function were dispersion free, its expectation value would be, contrary to its definition, representable as a linear combination of independent expectation values. Furthermore, if the ensembles are not homogeneous, then they are surely not dispersion free since homogeneity is easily shown to be a necessary though not sufficient condition for freedom from dispersion. Throughout the proof it is assumed that the quantities in question, and their corresponding operators, satisfy various quantum-mechanical relationships.

9. Notice that I am not making the trivial claim that the time dependence of the potential function $V = V(q,t)$ differs in each transition. This would be just what we expect. But mere change in the time dependence would not help, since the integral which must vanish is an integral over the coordinates —e.g., positions or momenta—and would be unaffected by a change in the time dependence of $V(q,t)$.

10. See Bohm (in Bates 1962) and Freistadt (1957) for detailed descriptions of various hidden variable theories.

11. Failure to fully grasp the cited distinction mars an otherwise valuable review article by Ballantine (1970).

12. Jauch (1968) purports to show that hidden variable theories are "empirically" false because they imply, contrary to the evidence supporting quantum mechanics, the *compatibility* of all "propositions" concerning the various properties of a given system at a given time. This is most misleading and unsatisfactory. The alleged *incompatibility* of certain pairs of propositions in quantum mechanics is based on an equivocation so obvious as to be embarrassing. Discussion of this point properly belongs in the next chapter, on quantum logic.

13. The man who introduced "instantaneous action at a distance" into physics was also the man who reminded us that anyone who knows "even a little philosophy" can see the absurd character of the expedient.

Chapter 5

1. My formulation of the interpretation under attack is based on the discussion of Putnam (1966–68). Though Putnam's discussion is relatively naive and unrigorous, his position is not essentially different from that of Jauch (1968), Finkelstein et al. (1962–63), or Birkhoff and von Neumann (1936). Even Tomonaga (1966) is infected by the confusion underlying this interpretation.

2. Heisenberg was always well aware of this point. See Heisenberg 1930, p. 20.

3. The reader should be prepared for the following query: given finite, not infinitesimal, slit widths, what is the agency which affects the particle's path? Answer: either the particle-slit edge interaction changes the particle's path or else it doesn't get changed. This answer will be followed by the query: but if the latter case (streaming) predominates overwhelmingly, how is the interference effect to be explained? Answer: if interaction is negligible compared to streaming, no interference pattern will be formed. It will take some time to convince the questioner his puzzlement stems from contradictory presuppositions, but it can be done.

Chapter 6

1. The label 'linear transport theory' is happier than the more standard label 'neutron transport theory' for two reasons. First, the theory applies to many kinds of particles other than neutrons. Second, mention of neutrons may lead to the mistaken conclusion that transport theory is, or should be, a branch of quantum mechanics rather than a branch of classical statistical mechanics. It is true that quantum mechanics is required for the accurate description, prediction, and explanation of the microscopic particle processes involved—scattering, absorption, fission, etc. But, strictly speaking, this fact is no part of the business of transport theory, which takes the probabilities—"cross sections"—for these processes as input data without inquiring into the origin of the data. The task of transport theory is to calculate particle population distributions, from this data, as functions of space, time, energy, angular distributions, etc.

2. It is worth noting that, strictly speaking, the Schrödinger equation does not have the form of a wave equation at all, but that of a diffusion equation.

3. Notice that if what is wanted in the way of "explanation" of the Schrödinger equation is its derivation from a more basic theory then this desire cannot be satisfied within the exercise we are engaged in—the interpretation of the present theory. It could only be satisfied by formulating and experimentally verifying the more basic theory. But at present we have no reason to believe any such "more basic" theory is true.

4. Once again, the reader is reminded that a demand for an "explanation" of the uncertainty relations—just as of the Schrödinger equation which contains these relations implicitly—in terms of a more basic theory is a demand which could only be satisfied by the discovery of a new physical theory, and not by a (philosophic and/or physical) interpretation of the present theory. Needless to say, new theories are not discoverable on demand.

5. See Tomonaga (1966), 2: chaps. 8, 9, for a lucid exposition of this and other aspects of the standard interpretation.

Notes

6. I emphasize that I am here only setting up the standard analysis in order to criticize it later.

7. The term 'EPR experiment'—for Einstein, Podolsky and Rosen, co-authors of the 1935 article under discussion—has become a generic label for any thought experiment which seems, "paradoxically," to show that simultaneous determination of values of conjugate variables is possible. Many recent EPR experiments discuss spin effects.

8. See the recently published Born-Einstein correspondence (Born 1971).

9. The famous Gibbs energy "paradox," on analogy with which Landé once justified a rederivation of the foundations of quantum mechanics—see Landé (1955), chapter 1—is not very paradoxical. If causes and effects are properly sorted out, no cause-effect discontinuity appears.

Chapter 7

1. This is the interpretation of, among many others, March (1951) and Mehlberg, "The Problem of Physical Reality in Contemporary Science," in Bunge (1967).

2. Even the seemingly intractable Klein-Gordon equation is susceptible to a consistent particle interpretation if we assume no fields strong enough for pair creation or annihilation. For then the positive and negative energy solutions of the Klein-Gordon equation are decoupled, and the probability density for a given particle involves only one kind of wave function solution and is always non-negative.

3. If we introduce considerations from *general* relativity, we might claim there is some indication that gravitational fields are radiation fields; but there is no basis for such a claim if we limit the discussion to *special* relativity plus classical mechanics.

4. If we insist on calling photons "particles" because discrete energy and momentum exchanges are involved in field-field and field-particle inter-actions, we may as well say our economic system is a system of "particles" because currency exchanges take place in discrete units. In some special contexts, this metaphor might be useful, but in general it is more misleading than informative.

Bibliographic Note

This somewhat polemical note is intended to provide a useful introductory guide to a small, but representative, sample of the vast and abstruse literature on quantum mechanics and related topics in philosophy of science, and a supplement to the meager list of specified references in the body of the text.

Jammer's history of nonrelativistic quantum mechanics is invaluable as a complete documentary of original source material—books, articles, papers, letters, even private conversations—through 1964. Tomonaga's historically oriented text is the single most useful book on quantum theory. Though completely rigorous, the mathematical derivations are always preceded and followed by intelligible English explanations of their physical significance. Van der Waerden and Whittaker are helpful on various specific points.

My exposition of the Copenhagen interpretation is based on Bohr's 1927 Solvay Congress Paper (in Bohr 1961), Heisenberg's 1929 Chicago lectures (Heisenberg 1930), and Bohr's 1935 reply (Bohr 1935) to the thought experiment of Einstein, Podolsky, and Rosen (1935). Hanson (1963) sometimes recognizes the tension between the correspondence principle and the claim of conceptual incompatibility between classical and quantum theory, and proposes to avoid inconsistency by reinterpreting the principle as a trivial and uninformative claim about the relation of two formal calculi. Feyerabend (1968–69), following Meyer-Abich very closely, interprets Bohr as saying macroscopic systems, though not atomic ones, have exact, stable, intrinsic properties independent of interaction with other systems. How this can be squared with Bohr's flat statements to the contrary is a great mystery.

Indeed, the Meyer-Abich/Feyerabend interpretation depends on two groups of citations from Bohr's writings: those like the forementioned, which cry out for the opposite interpretation, and those which are irrelevant because concerned with the old, semiclassical Bohr-Sommerfeld theory. To pass off the latter as relevant, Meyer-Abich and Jammer, with Feyerabend's approval, are literally driven to flights of fancy—to saying that Bohr's Copenhagen interpretation sprang fullblown into his mind upon reading a passage from (William) James that describes birds hopping from limb (stationary state) to limb (stationary state) with no resting places in between. To picture Bohr as a man with a preconceived solution in hand, waiting for a problem to be solved, does no justice to his careful reasoning from particular theoretical and experimental discoveries. It is ironic that these same writers have also emphasized this "careful empiricist" aspect of Bohr's character. Jammer, Feyerabend, and Meyer-Abich are not men who allow themselves to be hamstrung by consistency.

The Copenhagen interpretation was not infected with subjectivist elements in its formative and influential years, 1927–35, but ostensibly subjectivist passages can be found in the later writings of Heisenberg, especially *Physics and Philosophy* (1958), and in some (1965), though not the most recent (1968) of the writings of declared Copenhagen adherent Rosenfeld.

I rest no weight on standard criticisms of the positivist verification theory of meaning, or of operationalism, an early, abortive offshoot of positivism, since the 1927–35 views of Bohr and Heisenberg do not presuppose these dubious tenets. Still, the criticisms, and the views criticized, are interesting in their own right. The verification theory of meaning evolved, beginning in the 1920s, in the writings of such positivists as Schlick, Carnap, Ramsey, and Neurath. A classic statement of the theory can be found in Ayer (1952), and a summary of early difficulties arising from attempts to formulate the theory precisely is given in Hempel (1950). Subsequently, a more elaborate positivist theory was developed, partly in response to these difficulties. Various versions are given by Carnap (1936–37, 1956), Hempel (1958), Nagel (1961), and Braithwaite (1968); all involve a radical distinction between theoretical terms and observation terms, and the notion of "partial interpretation" of the theoretical terms. This neopositivist apparatus has been criticized by Achinstein (1968), Rozeboom (1960), Spector (1965–66), Shapere (1965), Hesse (1952, 1953, 1958, 1962), Putnam (1962a), Maxwell (1962), Suppe (1972), and countless others.

Some of these critics argue that partial interpretation cannot achieve its primary goal of distinguishing empirical science from other, allegedly meaningless, discourse—and/or that it is not necessary for this purpose. Some argue that the theory/observation distinction is fundamentally unsound, reflecting only transient, pragmatic limits on observation and obscuring other, far more important epistemic and metaphysical distinctions. Some critics use both kinds of arguments. While I find both arguments cogent, I rest no weight on either, preferring to argue that, even given his (idiosyncratic) version of the positivist apparatus, Nagel cannot show what he must, to defend Bohr and attack me: that a certain group of terms—'position of an electron', 'momentum of an electron', etc.—are theory-dependent in meaning.

Feyerabend's thesis, that the meanings of all scientific terms are theory-dependent, is expounded in (1962a, 1962b, 1965), and criticized by Achinstein (1964) and Shapere (1966). Further discussion and criticism is given in Wartofsky (1965).

The classic statement of operationalism is given by Bridgman (1927). Many social scientists, including Watson (1924), Skinner (1953), and Chapin (1939), have subscribed to this position, which is criticized in Lindsay (1937) and Hempel (1954, 1966).

The Margenau-Nagel deterministic construal of quantum mechanics was first developed by the former as early as 1934. Margenau has also written on the quantum theory of measurement (1958, 1963). His distinction between *measurement* (specification, usually by after-the-fact calculation, of the magnitude of a property) and *state-preparation* (bringing about, by causal interaction, the state of the physical system to be measured) is helpful in understanding the "pinhole," EPR, and double-slit thought-experiments. Passage through the pinhole, or one of the slits, constitutes state-preparation; later interaction with a detector constitutes "state-destruction"; measurement is, strictly speaking, an inference from the observed results of those two processes. Margenau criticizes the theories of measurement, loosely associated with the Copenhagen interpretation, developed by von Neumann (1955), London and

Bauer (1939), and Wigner (1963, 1967). If one applies a wave functional-Hamiltonian description of the total system—macroscopic instruments plus, say, microscopic particle—involved in typical measurements, one finds that it seems impossible to attribute exact, stable, intrinsic properties to the macroscopic objects involved.

The cited authors, among many others, try to get around this difficulty by assuming a sudden, discontinuous, unpredictable and uncontrollable change in the system upon measurement-interaction. (This is the so-called *projection postulate*; the process allegedly involved is called "reduction" of the wave packet, or of the state vector.) This move does not remove the difficulty, which is not that the particular final state of the system is not predictable, but rather that what seems to be predicted is that the system will not be in any particular state at all—e.g., the "pointer" on the meter won't point anywhere in particular. In desperation, von Neumann introduced a third component for the particle-instrument system to interact with, leaving the latter system, though not the new, three-component system, in a definite state. In actual experiments this extra "interaction" could only be the perceptual experience of the experimenter, and so Wigner has been led to say that the consciousness of the observer affects the physical processes involved!

It is crucial that we realize this measurement problem is created entirely by the assumption, often tacit, that the wave function describes individual processes, rather than statistical ensembles thereof. We have seen, ad nauseam, why this assumption cannot be true. Once we accept Bohr's sound insight that particular processes are indeterministic, and reject his (usually tacit) assumption that the wave function describes individual processes, no problem arises. A discussion of these points is given in the review article (1970) by Ballantine, who, unfortunately, does not realize that quantum theory is essentially indeterministic. That is, he fails to distinguish the truth of the matter (that the theory is both statistical, "ensemble-describing," and indeterministic) from the hidden variable thesis (that the theory is statistical, but really deterministic).

It is this confusion which underlies current feverish discussions as to whether quantum mechanics can be "extended to a classical theory"—see Shewell (1959), Moyal (1949), Margenau and Hill (1961), Cohen (1966a), Margenau and Cohen (1967), Bopp

(1956, 1957), and Prugovecki (1967)—and whether, in the light of von Neumann's theorem, and recent elaborations of it, hidden variable interpretations of quantum theory are possible—see de Broglie (1953, 1960), Weizel (1953), Janossy (1953), Bohm and Bub (1966), Bates (1962), Freistadt (1957), Blokhintsev (1953, 1968), Alexandrow (1952), Bopp (1953), Feynes (1952), Ballantine (1970), Bell (1964, 1966), Gleason (1957), Jauch and Piron (1963), and Kochen and Specker (1967), Gudder (1970), Friedman and Glymour (1972), and Yourgrau and van der Merwe (1971). Similar confusion infects analyses of deterministic theories by Montague (1962) and Earman (1971).

In discussing criteria for judging the confirmation of competing hypotheses, I take confirmation, or inductive probability, to be a matter of rational belief, and proceed accordingly. That is, I follow Carnap (1950, 1952, 1963) in distinguishing sharply between inductive and statistical probability—roughly, relative frequency—and in holding a logical interpretation of the former. My version is, however, qualitative, like that of Strawson (1952), not quantitative, like Carnap's. Alternative views of confirmation are held by the subjectivists Ramsey (1931), Savage (1954), Jeffrey (1965), and Good (1962); by Ryle (1957), Toulmin (1958), and Harré (1957), who claim that theoretical or hypothetical beliefs are not involved at all, in making inductive inferences; by Popper (1959, 1962), who would have us abandon the notion of confirmation, or inductive probability, entirely; by pragmatic theorists like von Mises (1957), Reichenbach (1949), Feigl (1963), and Salmon (1963), who regard confirmational judgments not as complex statements, true or false, but as methodological prescriptions for action. Surveys of these and other theories of confirmation can be found in Kyburg (1970), including discussion of the crucial notion of simplicity. I emphasize again that all formal characterizations of simplicity are misguided, and hence all attacks on such characterizations are irrelevant to the nonformal notion of simplicity I rely on. Cf. Barker (1957), Goodman (1958–59), Kemeny (1955), Svenonius (1955), Kyburg (1961), and Ackermann (1962).

As for revisions of logic inspired by quantum mechanics, Heelan (1970) has pointed out difficulties common to those—e.g., Birkhoff and von Neumann (1936), Finkelstein et al. (1962–63), Finkelstein (1962–63, 1968), Jauch (1968), and Putnam (1966–68)—who claim that the distributive laws of logic fail in the "empirically correct quantum logic." Such claims are often self-referentially inconsistent, for the very forms of inference held to be empirically refuted are used in defense of the claims. Further, none of the cited authors distinguish among the different kinds of statements employed in quantum theory, misleadingly lumping them all together as "basic empirical propositions." If these relatively minor revisions of logic are unjustified, more drastic revisions—rejection of the law of excluded middle, multivalued logics, etc.—are unworthy of the consideration urged by Reichenbach (1946), von Weizäcker (1955), and others.

Fine (1968) manages, with less than no awareness of the physics involved, to make the valuable points that the claim that "quantum logic" is nonclassical falsely presupposes that the only (sic) way of formulating probability statements in quantum theory employs a non-Boolean propositional calculus; and that the claims that quantum mechanics presupposes a nonclassical probability calculus are incoherent. See Suppes (1966) for a contrary position, and Mackey (1963), Cohen (1966b), and the previously cited works of Margenau (1963) and Moyal (1949) for other technical analyses of the probabilistic aspects of quantum theory.

Schwinger's anthology (1958) contains all the most important of the classic technical papers on relativistic quantum mechanics and quantum field theory, including the work of Dirac, Heisenberg, Jordan, Klein, Wigner, Dyson, Tomonaga, and many others. Sakurai (1967) is the most useful text on such advanced topics, but Jauch and Rohrlich (1955), Bjorken and Drell (1968), Roman (1960), and Schiff (1968) are helpful on specific points. See Margenau (1944) for odd and interesting claims about the philosophical significance of the Pauli exclusion principle.

Finally, useful summaries, by several of the participants, of the quarter-century debates among Bohr, Born, Einstein, Pauli, and others can be found in Schilpp (1970).

References

Achinstein, P. A. 1964. *J. Phil.* 61: 701.

———. 1968. *Concepts of Science.* Baltimore: Johns Hopkins Univ. Press.

———. 1969. In *The Business of Reason*, ed. J. J. Macintosh and S. C. Coval, pp. 1–25. London: Routledge and Kegan Paul.

Ackermann, R. 1962. *Phil. Rev.* 71: 236–40.

Alexandrow, A. 1952. *Dokl. Akad. Nauk.* 84.

Ayer, A. J. 1952. *Language, Truth and Logic.* New York: Dover.

Ballentine, L. E. 1970. *Rev. Mod. Phys.* 42: 358–81.

Barker, S. F. 1957. *Induction and Hypothesis.* Ithaca: Cornell Univ. Press.

Bates, D. R., ed. 1961. *Quantum Theory.* New York: Academic Press.

Becker, R. A. 1954. *Introduction to Theoretical Mechanics.* New York: McGraw-Hill.

Bell, J. S. 1964. *Phys.* 1: 195–200.

———. 1966. *Rev. Mod. Phys.* 38: 447–52.

Birkhoff, G., and von Neumann, J. 1936. *Ann. Math.* 37: 823–43.

Bjorken, J. D., and Drell, S. D. 1968. *Relativistic Quantum Fields.* New York: McGraw-Hill.

Blokhintsev, D. I. 1953. *Sowjetwissenschaft.* 6.

———. 1968. *The Philosophy of Quantum Mechanics.* New York: Humanities Press.

Bohm, D. 1952. *Phys. Rev.* 85: 166–93.

Bohm, D., and Bub, J. 1966. *Rev. Mod. Phys.* 38: 453–69.

Bohr, N. 1935. *Phys. Rev.* 48: 696–702.

———. 1961. *Atomic Theory and the Description of Nature*, pp. 52–91, 102–119. Cambridge: Cambridge Univ. Press.

Bohr, N., Kramers, H. A., and Slater, J. C. 1924. *Phil. Mag.* 47: 785–802.

Bopp, F. 1947. *Z. Naturforsch.* 2a: 202–16.

———. 1952. *Z. Naturforsch.* 7a: 82–87.

————. 1956. *Ann. L'Inst. H. Poincaré.* 15: 81–112.

————. 1957. *Observations and Interpretations,* ed. S. Körner and M. H. L. Price, pp. 189–96. London: Butterworth.

Born, M. 1956. *Physics in My Generation.* London: Pergamon Press.

————. 1964. *Natural Philosophy of Cause and Chance.* New York: Dover.

————, ed. 1971. *The Born-Einstein Letters.* New York: Walker.

Braithwaite, R. B. 1968. *Scientific Explanation.* Cambridge: Cambridge Univ. Press.

Bridgman, P. W. 1927. *The Logic of Modern Physics.* New York: Macmillan.

Bunge, M., ed. 1967. *Quantum Theory and Reality.* New York: Springer-Verlag.

Carnap, R. 1936–37. *Phil. of Sci.* 3: 419–71; 4: 1–40.

————. 1950. *The Logical Foundations of Probability.* Chicago: Univ. of Chicago Press.

————. 1952. *The Continuum of Inductive Methods.* Chicago: Univ. of Chicago Press.

————. 1956. In *Minnesota Studies in the Philosophy of Science.* Vol. 1, ed. H. Feigl and G. Maxwell, pp. 38–76. Minneapolis: Univ. of Minnesota Press.

————. 1963. In *The Phliosophy of Rudolf Carnap,* ed. P. A. Schilpp, pp. 859–1013. LaSalle: Open Court.

Chapin, F. S. 1939. *Soc. Forces.* 18: 153–60.

Cohen, L. 1966a. *J. Math. Phys.* 7: 781–86.

————. 1966b. *Phil. of Sci.* 33: 317–22.

Corben, H. C., and Stehle, P. 1960. *Classical Mechanics.* New York: Wiley.

de Broglie, L. 1953. *La Physique Quantique restera-t-elle indéterministe?* Paris: Gauthier-Villars.

————. 1960. *Non-Linear Wave Mechanics.* New York: Elsevier.

Duane, W. 1923. *Proc. Nat. Acad. Sci.* 9: 158–64.

Earman, J. 1971. *J. Phil.* 68: 729–44.

Ehrenfest, P., and Epstein, P. S. 1924. *Proc. Nat. Acad. Sci.* 10: 133–39.

————. 1927. *Proc. Nat. Acad. Sci.* 13: 400–409.

Einstein, A., Podolsky, B., and Rosen, N. 1935. *Phys. Rev.* 47: 777–80.

Evans, J. L. 1953. *Mind.* 62: 1–19.

Feigl, H. 1963. In *Philosophical Analysis,* ed. M. Black, pp. 113–47. Englewood Cliffs: Prentice-Hall.

Feigl, H., and Maxwell, G., eds. 1961. *Current Issues in the Philosophy of Science.* New York: Holt, Rinehart.

Feyerabend, P. K. 1962a. In *University of Pittsburgh Series in Philosophy of Science.* Vol. 1, ed. R. G. Colodny, pp. 189–263. Pittsburgh: Univ. of Pittsburgh Press.

————. 1962b. In *Minnesota Studies in the Philosophy of Science.* Vol. 3, ed. H. Feigl and G. Maxwell, pp. 28–97. Minneapolis: Univ. of Minnesota Press.

References

————. 1965. In *University of Pittsburgh Series in Philosophy of Science*. Vol. 2, ed. R. G. Colodny, pp. 145–260. Englewood Cliffs: Prentice-Hall.

————. 1968–9. *Phil. of Sci.* 35: 309–31; 36: 82–105.

Feynes, I. 1952. *Z. Phys.* 132: 81–106.

Fine, A. 1968. *Phil. of Sci.* 35: 101–11.

Finkelstein, D. 1962–63. *Trans. New York Acad. Sci.* 25: 621–37.

————. 1968. *International Center for Theoretical Physics: Report Ic/68/35*. Trieste.

Finkelstein, D., et al. 1962–63. *J. Math. Phys.* 3: 207–20; 4: 136–40.

Freistadt, H. 1957. *Nuovo Cimento Suppl.* 5: 1–70.

Friedman, M., and Glymour, C. 1972. *J. of Phil. Logic.* 1: 16–28.

Gleason, A. M. 1957. *J. Math. and Mech.* 6: 885–93.

Good, I. J. 1962. *Logic, Methodology and Philosophy of Science*. Vol. 1, eds. E. Nagel, P. Suppes, and A. Tarski, pp. 319–29. Stanford: Stanford Univ. Press.

Goodman, N. 1958–59. *Phil. and Phenom. Res.* 19: 429–46.

Gudder, S. P. 1970. *J. Math. Phys.* 11: 431–36.

Hanson, N. R. 1958. *Patterns of Discovery*. Cambridge: Cambridge Univ. Press.

————. 1963. *The Concept of the Positron*. Cambridge: Cambridge Univ. Press.

————. 1967. *Encyclopedia of Philosophy*, ed. P. Edwards. Vol. 7, pp. 41–49. New York: Macmillan.

Harré, R. 1957. *Phil.* 32: 58–64.

Heelan, P. 1970. *Synthese.* 21: 2–33.

Heisenberg, W. 1930. *The Physical Principles of the Quantum Theory*. Chicago: Univ. of Chicago Press.

————. 1958. *Physics and Philosophy*. New York: Harper.

Hempel, C. G. 1950. *Rev. Int. de Phil.* 4: 41–63.

————. 1954. *Scientific Monthly.* 79: 215–20.

————. 1958. In *Minnesota Studies in the Philosophy of Science*. Vol. 2, ed. H. Feigl and G. Maxwell, pp. 37–98. Minnesota: Univ. of Minnesota Press.

————. 1966. *Philosophy of Natural Science*. Englewood Cliffs: Prentice-Hall.

Hesse, M. 1952. *Brit. J. Phil. Sci.* 2: 281–94.

————. 1953. *Brit. J. Phil. Sci.* 4: 198–214.

————. 1958. *Brit. J. Phil. Sci.* 9: 12–28.

————. 1962. *Forces and Fields*. New York: Nelson.

Jammer, M. 1966. *The Conceptual Development of Quantum Mechanics*. New York: McGraw-Hill.

Janossy, L. 1953. *Ann. Physik.* 11: 323–61.

Jauch, J. M. 1968. *Foundations of Quantum Mechanics*. Reading: Addison-Wesley.

Jauch, J. M., and Rohrlich, F. 1955. *Theory of Photons and Electrons*. Cambridge: Addison-Wesley.

Jauch, J. M., and Piron, C. 1963. *Helvetia Phys. Acta.* 36: 827–37.

Jeffrey, R. 1965. *The Logic of Decision*. New York: McGraw-Hill.

Joos, G. 1934. *Theoretical Physics*. New York: Hafner.

Kemeny, J. G. 1955. *J. Phil.* 52: 722–33.

Kleene, S. C. 1952. *Introduction to Metamathematics*. New York: Van Nostrand.

Kochen, S., and Specker, E. P. 1967. *J. Math. and Mech.* 17: 59–87.

Kuhn, T. S. 1962. *The Structure of Scientific Revolutions*. Chicago: Univ. of Chicago Press.

Kyburg, H. E. 1961. *Phil. Rev.* 70: 390–95.

———. 1970. *Probability and Inductive Logic*. New York: Macmillan.

Landau, L. D., and Lifshitz, E. M. 1958. *Quantum Mechanics—Non-Relativistic Theory*. London: Pergamon Press.

Landau, L. D., Akhiezer, A. I., and Lifshitz, E. M. 1967. *General Physics*. New York: Pergamon.

Landé, A. 1955. *Foundations of Quantum Theory*. New Haven: Yale Univ. Press.

———. 1960. *From Dualism to Unity in Quantum Mechanics*. Cambridge: Cambridge Univ. Press.

———. 1965. *New Foundations of Quantum Mechanics*. Cambridge: Cambridge Univ. Press.

Lindsay, R. B. 1937. *Phil. of Sci.* 4: 456–70.

London, F., and Bauer, E. 1939. *La Théorie de L'Observation en Quantum Mechanics*. Paris: Hermann et Cie.

Mackey, G. W. 1963. *The Mathematical Foundations of Quantum Mechanics*. New York: Benjamin.

Mandl, F. 1957. *Quantum Mechanics*. London: Butterworth.

March, A. 1951. *Quantum Mechanics of Particles and Wave Fields*. New York: Wiley.

Margenau, H. 1934. *Phil. of Sci.* 1: 133–48.

———. 1944. *Phil. of Sci.* 11: 187–208.

———. 1950. *The Nature of Physical Reality*. New York: McGraw-Hill.

———. 1958. *Phil. of Sci.* 25: 23–33.

———. 1963. *Ann. of Phys.* 23: 469–85.

Margenau, H., and Hill, R. N. 1961. *Prog. Theor. Phys.* 26: 722–38.

Margenau, H., and Cohen, L. 1967. In *Quantum Theory and Reality*, ed., M. Bunge, pp. 71–89. New York: Springer-Verlag.

Maxwell, G. 1962. In *Minnesota Studies in the Philosophy of Science*. Vol. 3, ed. H. Feigl and G. Maxwell, pp. 3–27. Minneapolis: Univ. of Minnesota Press.

Meyer-Abich, K. M. 1965. *Korrespondenz, Individualitat und Komplementaritat*. Weisbaden: Steiner.

Montague, R. 1962. In *Decisions, Values and Groups*, ed. N. F. Washburne, pp. 325–70. New York: Pergamon Press.

Moyal, J. E. 1949. *Proc. Cambr. Phil. Soc.* 45: 99–124.

Nagel, E. 1961. *The Structure of Science*. New York: Harcourt.

Pauli, W., ed. 1955. *Neils Bohr and the Development of Modern Physics*. New York: McGraw-Hill.

Popper, K. R. 1959. *The Logic of Scientific Discovery*. New York: Basic Books.

References

————. 1962. *Conjectures and Refutations*. New York: Basic Books.

Prugovecki, E. 1967. *Can. J. Phys.* 45: 2173–2219.

Putnam, H. 1962a. In *Logic, Methodology and Philosophy of Science*, ed. E. Nagel, P. Suppes, and A. Tarski, pp. 240–51. Stanford: Stanford Univ. Press.

————. 1962b. In *University of Pittsburgh Series in Philosophy of Science*. Vol. 1, ed. R. G. Colodny, pp. 75–101. Pittsburgh: Univ. of Pittsburgh Press.

————. 1966–68. In *Boston Studies in Philosophy of Science*. Vol. 5, ed. R. S. Cohen and M. Wartofsky, pp. 216–41. Dordrecht: Reidel.

Ramsey, F. P. 1931. *Foundations of Mathematics and Other Logical Essays*. New York: Harcourt.

Reichenbach, H. 1946. *Philosophical Foundations of Quantum Mechanics*. Los Angeles: Univ. of California Press.

————. 1949. *Theory of Probability*. Los Angeles: Univ. of California Press.

Roman, P. 1960. *Theory of Elementary Particles*. Amsterdam: North Holland.

Rosenfeld, L. 1965. *Prog. Theor. Phys.* Suppl. Issue, Extra Number: 222–31.

————. 1968. *Nucl. Phys.* A108: 241–44.

Rozeboom, W. W. 1960. *Phil. Studies*. 11: 33–38.

Ryle, G. 1956. *Dilemmas*. Cambridge: Cambridge Univ. Press.

————. 1957. In *Observations and Interpretations*, ed. S. Körner and M. H. L. Price, pp. 165–88. London: Butterworth.

Sakurai, J. J. 1967. *Advanced Quantum Mechanics*. Reading: Addison-Wesley.

Salmon, W. 1963. In *Induction*, ed. H. Kyburg and E. Nagel, pp. 27–41. Middleton: Wesleyan Univ. Press.

Savage, L. J. 1954. *The Foundations of Statistics*. New York: Wiley.

Schiff, L. I. 1968. *Quantum Mechanics*. New York: McGraw-Hill.

Schilpp, P. A., ed. 1970. *Albert Einstein: Philosopher-Scientist*. LaSalle: Open Court.

Schrödinger, E. 1935. *Science and the Human Temperment*. New York: Dover.

————. 1952. *Brit. J. Phil. Sci.* 3: 109–23; 233–42.

Schwinger, J., ed. 1958. *Selected Papers on Quantum Electrodynamics*. New York: Dover.

Shapere, D. 1966. In *University of Pittsburgh Series in Philosophy of Science*. Vol. 3, ed. R. G. Colodny, pp. 41–85. Pittsburgh: Univ. of Pittsburgh Press.

————. 1965. *Philosophical Problems of Natural Science*. New York: Macmillan.

Shewell, J. R. 1959. *Amr. J. Phys.* 27: 16–21.

Skinner, B. F. 1953. *Science and Human Behavior*. New York: Macmillan.

Slater, J. C. 1968. *Quantum Theory of Matter*. New York: McGraw-Hill.

Spector, M. 1965–66. *Brit. J. Phil. Sci.* 16: 121–42; 17: 1–20.

REFERENCES

Strawson, P. F. 1952. *Introduction to Logical Theory.* New York: Wiley.

Suppe, F. 1972. *Phil. of Sci.* 39: 1–19.

Suppes, P. 1966. *Phil. of Sci.* 33: 14–21.

Svenonius, L. 1955. *J. Symbolic Logic.* 20: 235–50.

Tomonaga, S. 1966. *Quantum Mechanics.* New York: Interscience.

Toulmin, S. 1958. *The Uses of Argument.* Cambridge: Cambridge Univ. Press.

Van der Waerden, B. L. 1967. *Sources of Quantum Mechanics.* Amsterdam: North Holland.

Von Mises, R. 1957. *Probability, Statistics and Truth.* New York: Macmillan.

Von Neumann, J. 1955. *Mathematical Foundations of Quantum Mechanics.* Princeton: Princeton Univ. Press.

Von Weizäcker, C. F. 1955. *Naturwiss.* 42: 521–29; 545–55.

Wartofsky, M., ed. 1965. *Boston Studies in Philosophy of Science.* Vol. 2, pp. 157–262. New York: Humanities Press.

Watson, J. B. 1924. *Behaviorism.* New York: Norris.

Weizel, W. 1953. *Z. Phys.* 134: 264–85; 135: 270–73.

Whittaker, E. T. 1951. *A History of the Theories of Aether and Electricity.* New York: Nelson.

Wigner, E. P. 1963. *Amr. J. Phys.* 31: 6–15.

———. 1967. *Symmetries and Reflections.* Bloomington: Univ. of Indiana Press.

Yourgrau, W., and van der Merwe, A. 1971. *Perspectives in Quantum Theory.* Cambridge: M. I. T. Press.

Index

197